BREAK
THROUGH

WHY WE CAN'T LE...

TO ENVIRONMENTA...

Ted Nordhaus and
Michael Shellenberger

MARINER BOOKS

HOUGHTON MIFFLIN HARCOU...

BOSTON • NEW YORK

We dedicate this book to our parents,
Robert and Jean Nordhaus
Nancy and Don O'Brian
Bob Shellenberger and Judith Green

and to Robert J. Nordhaus

First Mariner Books edition 2009
Copyright © 2007 by Ted Nordhaus and Michael Shellenberger

For information about permission to reproduce selections from this book,
write to Permissions, Houghton Mifflin Harcourt Publishing Company,
215 Park Avenue South, New York, New York 10003.

www.hmhbooks.com

Library of Congress Cataloging-in-Publication Data

Nordhaus, Ted.
Break through : from the death of environmentalism to the
politics of possibility / Ted Nordhaus and Michael Shellenberger.
p. cm.
Includes bibliographical references and index.
ISBN 978-0-618-65825-1
1. Environmentalism — Political aspects — United States.
2. Environmental policy — United States. 3. Political ecology —
United States. 4. United States — Politics and government.
I. Shellenberger, Michael. II. Title.
GE197.N67 2007   333.72 — dc22   2007018307

ISBN 0-978-0-547-08595-1 (pbk.)

Printed in the United States of America

DOC 10 9 8 7 6 5 4 3 2 1

Excerpts from "I Have a Dream" are reprinted by arrangement with
The Heirs to the Estate of Martin Luther King Jr., c/o Writers House
as agent for the proprietor, New York, NY. Copyright 1963 Martin
Luther King Jr.; copyright renewed 1991 Coretta Scott King.

CONTENTS

PREFACE TO THE
MARINER EDITION

*B*REAK THROUGH was published in early October 2007, the same week that Al Gore won the Nobel Peace Prize. Many pundits understandably wondered why we were still claiming, three years after we wrote the essay "The Death of Environmentalism," that environmentalism needed to die so that a new politics, one capable of dealing with global warming, could be born. While such strong medicine seemed appropriate in 2004, many of our fellow greens and progressives viewed it as unnecessary at best, counterproductive at worst, in 2007.

But in the months after *Break Through* was published, it became increasingly clear that the fundamental political economy of climate change had not shifted. We had argued that while modern environmentalism had succeeded in addressing problems such as air and water pollution by very modestly increasing the costs of modern life in prosperous societies, addressing climate change would fail if its success depended on establishing far higher energy costs for a global population that was much poorer.

The next year proved this argument (in the words of *Time* magazine) to be "prescient." In December 2007, China, India, and most of the rest of the developing world once again rejected the enactment of binding pollution limits at a United Nations meeting, asserting, as they had for over a decade, that slowing development for the global poor was an unacceptable price to pay to mitigate a problem that had been largely caused by the global rich. A few months later, the U.S. Senate again rejected cap-and-trade climate legislation, demonstrating that even affluent societies would be resistant to paying significantly more for energy in order to slow global warming. And even that eco-exemplar Europe, which had agreed to binding limits on carbon emissions and had established policies to make dirty energy sources like coal more expensive, was on a coal building boom, demonstrating that regulation-based market mechanisms, like cap-and-trade, would be insufficient to make expensive alternative energy technologies practical replacements for fossil fuels.

In retrospect, the extraordinary burst of media attention and public concern, particularly among elites, about global warming looks more like the end of an era than its pinnacle. Environmentalism, rather than overwhelming its opponents, ran aground on its own contradictions. A politics based on a suspicion of modernity and development could offer no answers for the nearly seven billion people around the world eager for both. Western environmentalists who moralized against the excesses of materialism found few followers among a global population more interested in achieving comfort and prosperity, or even among their own countrymen, who, unsurprisingly, found more anxiety in their $4-per-gallon gasoline and shrinking 401(k)'s than in rising sea levels and drowning polar bears. And the consensus among climate scientists over the catastrophic implications of rising carbon emissions ran headlong into the sobering consensus among energy experts that the technologies necessary to stabilize global carbon emissions were far from deployable on a meaningful scale or at a commercially viable cost.

Yet at the moment that the old environmentalism was once again demonstrating its incapacity to effect change, a new, post-environmental politics was rising in its place. Barack Obama, his candidacy flagging, proposed a $150 billion clean energy investment program that would become a defining element of his campaign and the top priority of his presidency. Gore, who had made no mention of technology, or any significant role for governments in procuring it, in either *An Inconvenient Truth* or his Nobel acceptance speech, reversed himself in the fall of 2008, calling for massive public investment in clean energy technologies in order to transform the global energy economy. Even green groups now peddle cap-and-trade policies as a mechanism to pay for President Obama's green stimulus and investment programs.

We describe this politics as post-environmental because the case for ecologically beneficial goals, such as building a clean energy economy, has become nonenvironmental. In the past, economic or national security arguments for ecological action were ancillary at best. Today they provide the central rationalization for ecological action. This shift goes beyond rhetoric to policy agendas and political priorities, and, as such, signifies the death of environmentalism.

Whether greens succeed in passing cap-and-trade legislation in the next Congress, or in rescuing the Kyoto framework from irrelevance at talks in Copenhagen in December 2009, efforts to regulate and reduce carbon will hinge upon the emergence of low-cost low-carbon technologies to replace our existing fossil fuel infrastructure. No regulatory, pricing, or market mechanism will succeed without a dramatic drop in the cost of alternative technologies. There is today a growing consensus that only governments can make the kinds of long-term investments in research, development, deployment, and infrastructure that would make such a transformation possible. Whatever name it goes by, and whether its advocates call themselves greens, environmentalists, or something else, this will be the work of the new, post-environmental politics.

We wrote *Break Through* with the hope of providing a conceptual and philosophical framework for this new politics. That framework extends well beyond the traditional topics of environmental concern. The new ecological crises are inextricably intertwined with the basic workings of the global economy, and no ecological politics can succeed that does not come to terms with the realities of development and globalization. In this new political era, efforts to preserve the Amazon will become inseparable from Brazil's efforts to develop the Amazon. Similarly, environmental-justice advocates are as likely today to see themselves as housing advocates, job counselors, and community development experts as they are to spend their time protesting toxic waste or diesel-fueled buses.

In the United States, the past two decades brought us record levels of both economic growth and social inequality. Even in the midst of the worst economic downturn in half a century, we are still a remarkably affluent society. But two decades of market worship and underinvestment in public infrastructure have left us ill prepared, intellectually, politically, and economically, to make the kinds of sustained investments in our future that have always marked American greatness.

And yet America has a remarkable capacity to remake itself on the fly, and we see promising signs that this renovation of the American dream may already be under way. Having elected a mixed-race president just a few years after many observers, both here and abroad, had written off the American public as hopelessly retrograde, America is on the verge of launching one of the largest public investment programs since the 1930s and 1940s. That program will have many elements, from health care and other social programs to traditional public works projects, but it appears that investment in the clean energy technologies and industries of the future will be a major feature.

To the degree that this is so, much of Obama's economic program may ultimately resemble FDR's war effort more than his New Deal. It was the war effort that pulled America out of the Great Depression and laid the economic foundation for twenty-five years of sustained and equitable economic growth after World War II. Notably, that effort involved the direct government procurement of technologically advanced manufactured goods. Back then it was tanks and airplanes; today it will be windmills, solar panels, hybrid cars, and technologies to capture carbon from smokestacks and even the ambient air.

The subtitle of this edition of *Break Through*—*Why We Can't Leave Saving the Planet to Environmentalists*—reflects our belief that the shift we first described, rather prospectively, in "The Death of Environmentalism" is well on its way to being realized. When *Break Through* was first published many critics accused us of creating a straw man of environmentalism—despite the fact that we had written about central figures in the green canon, from Rachel Carson and Paul Ehrlich to Al Gore and Jared Diamond. Perhaps the reason these environmentalist exemplars seemed so alien is because modern green readers have experienced a shift in their own values, one that has occurred gradually over the past decade. It may be that we are all now, to one degree or another, post-environmentalists.

FROM THE NIGHTMARE
TO THE DREAM

THIS BOOK was born from an essay, "The Death of Environmentalism: Global Warming Politics in a Post-Environmental World," that we wrote in the fall of 2004. We released the essay in pamphlet form at the annual conference of environmental donors and grantees, hoping to spark a conversation among insiders. What we didn't expect was that it would be read and debated by such a diverse audience, from college students to corporate executives, everywhere from Italy to Colombia to Japan, or that it would become a projection screen for the hopes and anxieties of the broader progressive community in the United States.

After all was said and done, the passages of our essay that seemed to resonate the most with readers were those that criticized environmentalists for their doomsday discourse. The most quoted lines in the essay were these:

> Martin Luther King Jr.'s "I have a dream" speech is famous because it put forward an inspiring, positive vision that carried a critique of the current moment within it. Imagine how history would have turned out had King given an "I have a nightmare" speech instead.

We went on to contrast the environmental movement's complaint-based approach to politics with King's positive vision — and called on environmentalists to replace their doomsday discourse with an imaginative, aspirational, and future-oriented one.

What we didn't know at the time we wrote those words was that King *had* given an "I have a nightmare" speech. In fact, he had given it just moments before he gave his "I have a dream" speech.

The setting was the August 28, 1963, March on Washington. Hundreds of thousands of people had crowded before the Lincoln Memorial, on the Washington Mall, to hear King and other leaders rally the country to support civil rights legislation. Millions of others watched on television, where the speech was carried live by all three networks.

President John F. Kennedy had just returned from Germany; against the backdrop of the Berlin Wall, he had called for freedom for those living behind the Iron Curtain. On his return, Kennedy asked King to call off the demonstration. "We want success in Congress," the president said, "not just a big show at the Capitol."[1]

Kennedy's comment tipped King into a dark mood. The worst manifestations of human nature were on display in the South — bigotry, beatings, cowardice, murder — and King was intent on making sure that white America, Kennedy included, faced up to them. And so, a few minutes before he was to speak, King leaned over to the gospel singer Mahalia Jackson, who had been traveling the country with him, and whispered, "Before I speak I want you to sing 'I Been 'Buked and I Been Scorned.'" When Jackson told her stage manager of King's request, he replied, "We need a song that's a little livelier than that!" But Jackson did as King requested. "Dere is trouble all over dis world, children," she sang. "Dere is trouble all over dis world."[2]

The operating metaphor in King's nightmare speech was the debt white America owed African Americans. "We've come to our nation's capital to cash a check," he said, but "instead of honoring this sacred obligation, America has given the Negro people a bad check, a check that has come back marked 'insufficient funds.'" The words revealed King's fears that the march wouldn't be taken seriously by

Congress and the White House. "It would be fatal for the nation to overlook the urgency of the moment," he warned. Those who underestimated the movement's power, he said, would have a "rude awakening."[3] It was perhaps the darkest and most discouraged speech King ever gave.

But then something strange and wonderful happened. A voice rang out from the back of the dais. It was Mahalia Jackson. "Tell them about your dream, Martin!" She could feel that King had dwelt too long in the dark valley — he needed to bring the crowd up to the sunlit mountaintop. Having heard him give riffs of the dream speech to earlier audiences, Jackson knew just what King needed to do. "Tell them about the dream!" she cried once more.[4]

King seemed to address his next line — "Let us not wallow in the valley of despair" — as much to himself as to the crowd. He then pattered — "I say to you today my friend" — and paused, triggering soft applause from the tired audience and buying himself the time he needed to reorganize his thoughts.

King then seemed to find the words Mahalia Jackson had tossed him, and he began the new speech. "And so even though we face the difficulties of today and tomorrow, *I still have a dream*."[5] From there King led the hot crowd in a rapid climb out of the valley.

> [W]hen we allow freedom to ring, when we let it ring from every village and every hamlet, from every state and every city, we will be able to speed up that day when all of God's children — black men and white men, Jews and Gentiles, Protestants and Catholics — will be able to join hands and sing in the words of the old Negro spiritual: "Free at last! Free at last! Thank God Almighty, we are free at last!"

With the words "Thank God Almighty, we are free at last," racial integration suddenly felt inevitable.

Even the nervous Kennedy, who was watching the speech live on television down the street, was impressed. "He's damn good," he told his aides. When Kennedy greeted King at the White House later that same day, the president smiled and said, "I have a dream."[6]

Three months later Kennedy was dead, but his successor, Lyndon Johnson, surprised nearly everyone and became an aggressive pursuer of King's dream. Over the next two years, Congress passed, and President Johnson signed, the sweeping Civil Rights and Voting Rights acts. And while those laws might have been enacted no matter what speech King had given, it is unlikely that history would have unfolded as peacefully or as quickly as it did had it not been for King's dramatic and mysterious leap from the nightmare to the dream.

1.

One unfortunate consequence of having quoted Dr. King in our essay was that we ended up tapping into that apparently inexhaustible reservoir of nostalgia for the 1960s. The truth is that King's dramatic leap from the nightmare to the dream can be a parable for the future only if we first understand how much the world has changed since 1963.

Schools today are still segregated, but for reasons vastly more complex than Jim Crow. Americans of all colors are living roughly a decade longer, thanks in part to advances in medicine. Our homes and cars are larger, and more of us own them. We take such luxuries as air conditioning, cell phones, and inexpensive air travel for granted. And our air and water are far cleaner. Our unprecedented wealth and freedom have profoundly changed what we care about, aspire to, and believe in, so it's no wonder that the old political and moral fault lines no longer apply.

Civil rights, the environment, feminism, labor — what were once cutting-edge movements are now established special interests. This is due in no small part to their success. Rights-based liberalism ended school segregation, dramatically reduced employment discrimination, and gave women the right to abortions. And environmental laws cleaned up our air and water and protected wilderness lands.

But the old politics has taken us as far as it can. The world has changed in profound ways, but liberal interest groups have not. In defining themselves and their interests so narrowly, it is the issue groups and their political allies — not bogeymen like Rush Limbaugh, Fox News, and the Heritage Foundation — who have created the widespread impression that liberalism is little more than an aggregation of the aggrieved. Environmentalists define their interest as limiting human intrusions upon nature. Health care reformers define theirs around insuring the uninsured. Civil rights groups define their mandate around ending racial prejudice and disparities. And reproductive rights and women's groups define theirs around access to contraception and abortion.[7]

This literalism in setting policy contains its own contradiction: in reducing their own manifold interests to single essential causes and complaints, liberal-issue groups have inhibited their ability to create the kinds of broad coalitions they need to achieve their goals. And in consistently defining the interests of others — whether they are corporate executives, labor unions, or Brazilian peasants — as outside the categories of the environment and nature, environmental and conservation leaders have failed to create a politics capable of dealing with ecological crises.

2.

In 1943, the American psychologist Abraham Maslow wrote a seminal paper called "A Theory of Human Motivation."[8] In it he introduced the theory that humans have a "hierarchy of needs," a deceptively simple concept that many of us can still remember seeing as a multicolor pyramid in our high school social studies classes. At the bottom of the pyramid there were the basic material needs: food, shelter, and security. Above those were esteem, belonging, and status needs, and above them were "being" needs, such as purpose, self-creation, and fulfillment. Maslow argued that once we have met the lower material needs, new and higher needs emerge that demand our attention. Sociologists today describe these higher needs as "postmaterial," since they emerge only after our basic material needs have been met.

Environmentalism and other progressive social movements of the 1960s were born of the prosperity of the postwar era and the widespread emergence of higher-order postmaterialist needs. As Americans became increasingly wealthy, secure, and optimistic, they started to care more about problems such as air and water pollution and the protection of the wilderness and open space. This powerful correlation between increasing affluence and the emergence of quality-of-life and fulfillment values has been documented in developed and undeveloped countries around the world.

Of course, these universal human *needs* express themselves as strikingly different social *values,* and we must thus understand both. By values we mean, very broadly, those fundamental concerns and beliefs that comprise our worldviews and the different ways of making sense of ourselves. An individual's core values are formed early in life and evolve slowly over time as we go through major life events. Throughout this book we will refer to the social values research that we and others have conducted.[9] Nothing is more central to this book than our contention that for any politics to succeed, it must swim with, not against, the currents of changing social values.

Environmentalists have long misunderstood, downplayed, or ignored the conditions for their own existence. They have tended to view economic growth as the *cause* but not the *solution* to ecological crisis. Environmentalists like to emphasize the ways in which the economy depends on ecology, but they often miss the ways in which thinking ecologically depends on prospering economically. Given that prosperity is the basis for ecological concern, our political goal must be to create a kind of prosperity that moves everyone up

Maslow's pyramid as quickly as possible while also achieving our ecological goals.

3.

Paradoxically, it is the global ecological crises themselves that have
triggered the death of environmentalism. For us to make sense of
them, the category of "the environment" — along with the ancient
story of humankind's fall from nature — is no longer useful. The
challenge of climate change is so massive, so global, and so complex
that it can be overcome only if we look beyond the issue categories of
the past and embrace a grand new vision for the future.

Through their stories, institutions, and policies, environmentalists constantly reinforce the sense that nature is something separate
from, and victimized by, humans. This paradigm defines ecological
problems as the inevitable consequence of humans violating nature.
Think of the verbs associated with environmentalism and conservation: "stop," "restrict," "reverse," "prevent," "regulate," and "constrain." All of them direct our thinking to stopping the bad, not creating the good. When environmentalists do speak in positive tones, it is
usually about things like clean air and water, or "preserving nature"
— all concepts that define human activity as an intrusion on, or a
contaminant of, a separate and once pure nature.

Environmental leaders continue to insist that global warming is
essentially a very big pollution problem. But while the coal smoke of
Manchester in nineteenth-century England and smog in Los Angeles
in the 1970s can reasonably be understood as pollution, the principal
greenhouse gas, carbon dioxide, is invisible and has no offensive
odor. Moreover, the quantitative accumulation of carbon dioxide in
the atmosphere has created something *qualitatively* different from
pollution: changing temperatures worldwide and melting ice caps,
which may lead to a collapse of the North Atlantic Gulf Stream, water
shortages, new disease epidemics, and resource wars.

To describe these challenges as problems of pollution is to stretch the meaning of the word beyond recognition. Global warming is as different from smog in Los Angeles as nuclear war is from gang violence. The ecological crises we face are more global, complex, and tied to the basic functioning of the economy than were the problems environmentalism was created to address forty years ago. Global warming threatens human civilization so fundamentally that it cannot be understood as a straightforward pollution problem, but instead as an existential one. Its impacts will be so enormous that it is better understood as a problem of *evolution,* not pollution.

Our planet, and we along with it, will evolve in rapid and dramatic ways over the next century. The challenge for humankind now is not whether we can stop global warming, which is already well under way, but whether we can minimize it, prepare for it, and improve human and nonhuman life while we're at it. The problem is so great that before answering *What is to be done?* we must first ask, *What kind of beings are we?* and *What can we become?*

4.

Before we wrote the essay "The Death of Environmentalism," the two of us had spent all of our professional careers, about thirty years between us, working for the country's largest environmental organizations and foundations, as well as many smaller grassroots ones. Like most of our colleagues, we viewed global warming as a problem of pollution, whose solution would be found in pollution limits.

In 2003 we started to break away from the pollution and regulation framework. With a small group of others we created a proposal for a new Apollo project. We proposed a major investment in clean-energy jobs, research and development, infrastructure, and transit, with the goal of achieving energy independence. The political thinking was that this agenda would win over blue-collar swing voters and Reagan Democrats in the presidential battleground states of the Midwest, and excite the high-tech creative class at the same time. And by putting serious public investment on the table — $300 billion over ten years — we hoped we could break through the logjam that had divided business, labor, and environmental groups for twenty years.

But more than any short-term political calculation, Apollo, we hoped, would be the vehicle for telling a powerful new story about American greatness, invention, and moral purpose.

After we created the Apollo proposal, we did what new political coalitions on the left tend to do: round up endorsements from other groups. And while we succeeded in getting endorsements and letters of support for Apollo's principles from businesses, unions, and most of the large national environmental groups, we were baffled, and then angered, by what happened next.

Environmental lobbyists told us that while they supported Apollo's vision, they would do nothing to support it in concrete ways, either in Congress or during the 2004 election. Those of us who had created Apollo had made the decision to focus on jobs and energy independence, because they were far higher priorities among voters than stopping global warming. In particular, we discovered that investment in clean-energy jobs, to get free of oil, was more popular with voters than talk of global warming, clean air, and regulation. But environmental leaders thought our nonenvironmental and nonregulatory focus was a vice, not a virtue.

Fearing that it would distract Democrats' attention away from stopping the George W. Bush administration's energy bill, which included billions in new subsidies for coal and oil, environmental leaders eventually asked us to keep Apollo legislation from being considered by Congress. Still the good soldiers, we did as we were asked, and Apollo was, briefly, withdrawn. But it hardly mattered: the Bush energy bill passed anyway.

5.

Frustrated with the environmental lobby's policy literalism, and annoyed by the uninspired, small-bore, complaint-based agenda of

Democratic presidential candidate John Kerry, we set out in the summer of 2004 to write an essay about the politics of global warming.

We started by interviewing the environmental leaders and funders who determine global warming strategy in the United States. By the time we finished, we were convinced that the environmental approach was inadequate, at the policy and the political levels, to deal with the monumental nature of the crisis. We concluded that the problem wasn't with environmental leaders so much as with their conceptual models, policy frameworks, and institutions.

The intensity of the reaction to our essay surprised, delighted, and occasionally frightened us. Many imagined that we had claimed environmentalism *was* dead. The response from the most literal-minded was that environmentalism couldn't be dead because they themselves were (a) environmentalists and (b) alive. Others didn't understand how we could be so concerned with global warming and *not* be environmentalists, implying that such a position was a contradiction in terms.

Happily, many people read the essay and, whether they agreed or disagreed, considered our thesis that "modern environmentalism, with all of its unexamined assumptions, outdated concepts and exhausted strategies, must die so that something new can live." Our intention was, in part, to question whether the category of "the environment" made sense any longer. If "the environment" *includes* humans, then everything is environmental and the concept has little use other than being a poor synonym for "everything." If it *excludes* humans, then it is scientifically specious, not to mention politically suicidal.

In the end, the most gratifying aspect of the experience was being told by environmentalists and nonenvironmentalists alike that the essay had had a powerful impact on their thinking and their work. Some told us that they read and discussed the essay in small groups of friends and colleagues. Local environmental leaders told us that they had become more focused on creating a new kind of development than on "protecting the environment."

Today, a new Apollo-like proposal for energy independence seems to appear every few months, including from the campaigns of presidential candidates. The story of America as an innovative nation, the increasing importance of high-tech research and development, and the role of strategic public investment have all emerged as key talking points for anyone concerned about global warming or energy independence. And billions of dollars in new investments are pouring into the clean-energy sector, and even major players in the old energy economy see the opportunity and are positioning themselves to take advantage of it. All of these are the makings of a new dream, and a new story, about America and the world.

6.

The political environment for action on energy independence and global warming has undergone a dramatic shift since 2004. Motivated by their anger with government inaction and the Bush administration's outright interference, climate scientists increasingly started speaking out about the need for bold action. In the summer of 2006, Al Gore wrote a best-selling book and starred in a widely seen movie, *An Inconvenient Truth,* that were compelling — and terrifying — presentations about global warming.

In lieu of action by Congress, progress on climate has come from other quarters. California enacted historic legislation reducing the state's greenhouse gases to 1990 levels by 2020, and other states are likely to follow. The U.S. Supreme Court ruled, in the spring of 2007, that the Clean Air Act gives the Environmental Protection Agency the authority to regulate carbon dioxide as a pollutant causing global warming. And sustainability is today one of the hottest topics in politics, the corporate world, and the media.

The twenty-year effort by environmentalists to educate the public about the facts of global warming has gotten us halfway there. Lawmakers and the media now understand the seriousness of climate change and are committed to action. The passage of federal legislation to cap greenhouse gas emissions, and create a mechanism for them to be traded, appears inevitable.

But if we are to seize the opportunities being offered, we must first face up to four inconvenient truths about global warming. The first is that those developed nations that ratified the Kyoto treaty on global warming have made little headway in actually reducing their own emissions. In late 2006, the United Nations announced that, since 2000, the emissions of the forty-one wealthy, industrialized members of Kyoto had gone up, not down, by more than 4 percent.[10]

The second truth is that China and India long ago rejected any approach to addressing climate change that would constrain their greenhouse gas emissions or their economic growth. For years, energy experts had expected that China would overtake the United States as the world's largest greenhouse gas emitter by 2025. It turns out that China will gain that dubious distinction by 2008.[11] The governments and the people of China and India are increasingly concerned about global warming, to be sure, but they are far more motivated by economic development, and to the extent that the battle against global warming is fought in terms of ecological limits rather than economic possibility, there's little doubt which path these countries will take.

The third truth is that even if we were to drastically limit the greenhouse gas emissions produced by power plants and automobiles, we would still need a strategy to slow the rapid rate of deforestation. Destruction of rain forests contributes an estimated 25 percent of all greenhouse gases, more than even vehicles contribute.[12] Perversely, some of the deforestation in Indonesia and Brazil is driven by the rising demand for land to grow biofuels.[13] In the years since the United Nations environmental conference in Rio de Janeiro in 1992, foreign governments and philanthropists have invested billions in conservation and "sustainable development" pilot projects in the Amazon region. And yet during that time, deforestation accelerated. The problem is confounded by the fact that when the forests are gone, they can no longer play their ecologically crucial roles of storing carbon and cooling the atmosphere.

The fourth truth about the crises we face is that global warming has arrived and will have increasingly serious consequences, even if we stop emitting all greenhouse gases tomorrow. Climatic changes will lead to increasingly severe, more destructive, and more deadly hurricanes, tornadoes, and monsoons. The melting of ice sheets will raise sea levels and increase the threat of flooding, agricultural collapse, and food shortages. In other parts of the world, global warming will likely trigger droughts, water scarcities, and famines.

7.

In November 2006, Americans voted to eject Republicans from both houses of Congress. Public upset over the worsening quagmire in Iraq has kept President Bush's popularity ratings at around 30 percent, and today it seems that nothing can go right for Republicans. The chances are good that in 2008 America will elect both a Democratic president and Congress, and so it is no exaggeration to say that the opportunity for real action on everything from global warming to health care is better than it has been since 1992.

Many on the left viewed the 2006 election results as proof not just that the Republican Party had been repudiated but also that conservative ideological hegemony had come to an end. Whether that's the case, only time will tell. What is certain is that, while voters rejected Republican incompetence, they have not yet affirmed a Democratic vision.

The time is ripe for the Democratic Party to embrace a new story about America, one focused more on aspiration than complaint, on assets than deficits, and on possibility than limits. For the party to do that, progressives, liberals, and Democrats must deal with some inconvenient truths of their own. Just as environmentalists must grapple with how global warming challenges the politics of limits, progressives must understand how a half century of prosperity and changing social values challenges materialist liberalism.

Globalization and the transition to a postindustrial economy have generated remarkable material wealth, but they have also brought outsourcing, downsizing, and instability. The result is that Americans have seen their wealth and spending power rise, but they have also become increasingly insecure in terms of their employment, retirement, health care, and community. What results is what we call insecure affluence, a kind of postmaterial insecurity that is profoundly misunderstood when viewed as poverty.

The worldview of materially affluent and postmaterially insecure people is vastly different from the worldview of the materially deprived. During the Great Depression, the poorest one-third of the country stood in breadlines, ate from the garbage, and roamed cities searching for work. They could not hide their poverty. Today's insecure affluent both mask and overcompensate for their insecurity by flaunting their material wealth. The politics born of material poverty cannot speak to postmaterial insecurity. Misunderstanding this, Democrats and liberals find themselves constantly telling Americans how poor and vulnerable they are — which is quite possibly the last thing insecure Americans want to be told.

The rise of insecure affluence has caused social values to evolve in two directions simultaneously. Rising insecurity has fueled the move away from fulfillment values and back toward lower-order, postmaterialist "survival" values, which tend to manifest as status competition, thrill-seeking, and hedonism, all of which have triggered a cultural backlash that conservatives more than liberals, Republicans more than Democrats, have harnessed. At the same time, rising affluence has fueled the shift, over the past century and a half, away from traditional forms of religious, familial, and political authority and toward greater individuality. In response, today Americans are creating a number of new identities for themselves that were unimaginable a hundred, fifty, or even twenty years ago. The problem is thus not with globalization and postindustrialization so much as with the absence of a new social contract, one that joins the individual's self-interest with the common good.[14]

8.

The politics we propose breaks with several widely accepted, largely unconscious distinctions, such as those between humans and nature, the community and the individual, and the government and the market. Few things have hampered environmentalism more than its longstanding position that limits to growth are the remedy for ecological crises. We argue for an explicitly pro-growth agenda that defines the kind of prosperity we believe is necessary to improve the quality of human life and to overcome ecological crises.

One of the places where this politics of possibility takes concrete form is at the intersection of investment and innovation. There is simply no way we can achieve an 80 percent reduction in greenhouse gas emissions without creating breakthrough technologies that do not pollute. This is not just our opinion but also that of the United Nations International Panel on Climate Change, of Nicholas Stern, the former chief economist of the World Bank, and of top energy experts worldwide. Unfortunately, as a result of twenty years of cuts in funding research and development in energy, we are still a long way from even beginning to create these breakthroughs.

The transition to a clean-energy economy should be modeled not on pollution control efforts, like the one on acid rain, but rather on past investments in infrastructure, such as railroads and highways, as well as on research and development — microchips, medicines, and the Internet, among other areas. This innovation-centered framework makes sense not only for the long-term expansion of individual freedom, possibility, and choice that characterize modern democratic nations, but also for the cultural peculiarities of the United States.

In 1840, Alexis de Tocqueville observed that "in the United

States, there is no limit to the inventiveness of man to discover ways of increasing wealth and to satisfy the public's needs." Rather than *limiting* the aspirations of Americans, we believe that we should harness them in order to, in Tocqueville's words, "make new discoveries to increase the general prosperity, which, when made, they pass eagerly to the mass of people."[15]

The good news is that, at the very moment when we find ourselves facing new problems, from global warming to postmaterialist insecurity, new social and economic forces are emerging to overcome them. The new high-tech businesses and the new creative class may become a political force for a new, postindustrial social contract and a new clean-energy economy.

One inspiring model for overcoming adversity can be found in the formation, after World War II, of what would later become the European Union. It was in the postwar years that the United States, France, Britain, and West Germany invested billions in the European Coal and Steel Community, which existed to rebuild war-torn nations and repair relations between former enemies, and which grew to become the greatest economic power the world has ever seen. Today's European Union wouldn't exist had it not been for a massive, shared global investment in energy. It's not hard to imagine what a similar approach to clean energy might do for countries like the United States, China, and India.

9.

Environmentalism offered something profoundly important to America and the world. It inspired an appreciation for, and an awe of the beauty and majesty of, the nonhuman world. It focused our attention on future generations and our responsibility toward them. And it called upon people to take valiant risks, from saving rain forests and whales to inventing wondrous new technologies that will help us overcome the ecological crises we face.

But environmentalism has also saddled us with the albatross we call the politics of limits, which seeks to constrain human ambition, aspiration, and power rather than unleash and direct them. In focusing attention so exclusively on the nonhuman worlds that have been lost rather than also on the astonishing human world that has been created, environmentalists have felt more resentment than gratitude for the efforts of those who came before us. And the "rational" environmentalist focus on just fixing what's wrong with the present narrows our vision at a time when we desperately need to expand it.

There are various expressions of environmentalism, and we have done our best to describe them with reference to specific events, leaders, and ideas. There are important differences between them, which we point out, but there are also striking similarities. When we describe the conceptual underpinnings these different expressions of environmentalism share, some readers will no doubt feel that we have overgeneralized. But those readers who do not see themselves in our descriptions and definitions of environmentalism might consider whether they themselves have already moved beyond environmentalism.

Others will wonder why, with our embrace of markets and prosperity, we don't just call the politics we are proposing a "new environmentalism." The reason has everything to do with our central argument: if we are to overcome ecological crises, we must no longer put concepts like nature or "the environment" at the center of our politics.

10.

When you really consider how monumental the ecological crises are, and how much they are an outgrowth of firmly rooted ways of being in the world, it is hard not to feel overwhelmed. And while fear is an appropriate response to crisis, it matters what we do with it. Fear may be inevitable, but despair is a choice.

With Abraham Lincoln at his back and Congress before him, Martin Luther King felt fear and resentment, and he expressed those dark feelings. But then he stopped himself midspeech. Perhaps he felt the crowd's wishes. Perhaps he heard Mahalia Jackson's cry. Perhaps he had scared himself sensible. Whatever the reason, consciously or unconsciously, King made a choice.

Today we have new choices to make. We must choose between a politics of limits and a politics of possibility; a focus on investment and assets and a focus on regulation and deficits; and a discourse of affluence and a discourse of insecurity. And, most of all, we must choose between a resentful narrative of tragedy and a grateful narrative of overcoming.

In the end, it was probably for the best that King gave a nightmare speech before giving the dream speech. Had he ignored his feelings of frustration and anger, his dream speech would not have been nearly as powerful. Had he avoided the dark valley, the mountaintop would not have been as high or as bright.

We will, to be sure, always call it the "I have a dream" speech. But we should never forget that it all began with a nightmare — one that King, and America with him, overcame.

PART ONE

THE POLITICS OF LIMITS

The greater part of what my neighbors call good I believe in my soul to be bad, and if I repent of any thing, it is very likely to be my good behavior. What demon possessed me that I behaved so well? You may say the wisest thing you can, old man — you who have lived seventy years, not without honor of a kind — I hear an irresistible voice which invites me away from all that. One generation abandons the enterprises of another like stranded vessels.

— HENRY DAVID THOREAU, *Walden*

1

THE BIRTH OF
ENVIRONMENTALISM

I N THE LATE 1960s, a new social movement swept through American
political life. Earlier that decade, Rachel Carson's Silent Spring woke
America to the dangers of pesticides. Smog was choking Los Angeles
and other cities. And in 1969, pollution on the surface of Cleveland's
Cuyahoga River burst into flames. That same year, Americans viewed
the first photograph of planet Earth taken from outer space and realized
how fragile and lonely our living planet truly is.

Modern environmental organizations emerged in response to these
newly visible consequences of industrialization. The Sierra Club transformed itself from a quiet hikers' club to a lobbying powerhouse. Two
years later, a group of young and idealistic attorneys determined to create an NAACP Legal Defense Fund for the environment founded the
Natural Resources Defense Council (NRDC) to bring the full weight of
scientific and legal expertise to bear on environmental policy. And on
April 22, 1970, twenty million Americans celebrated the first Earth Day.
The modern environmental movement was born, and it shone like a
candle in the dark night of race riots, political assassinations, and the
Vietnam War.

The new movement's impact on politics was swift and decisive.
Public outrage at these new pollution problems, combined with the environmental movement's deft use of science, lobbying, grassroots organizing, and the courts, led Congress to pass and presidents to sign dozens of

environmental policies into law, from the Clean Air Act to the Endangered Species Act. By the end of the 1970s, the United States had protected millions of acres of wilderness and public land, dramatically improved air and water quality throughout the nation, and established the strongest environmental protections of any nation on earth.

Or at least that's how the story goes. As far as political fables go, this genesis story has served the environmental movement well. It depicts environmental leaders as the parents of America's most important environmental laws. It defines public support for environmental action as a relatively simple reaction to visible pollution. It imagines Earth Day to have been a spontaneous grassroots expression of popular discontent. And it establishes scientific and legal expertise as the basis of the legislative victories of the era.

And while most of the facts commonly marshaled to tell the environmentalist birth story are technically correct, the overall narrative is all wrong.

1.

On June 22, 1969, oil and debris on the surface of the Cuyahoga River in Cleveland, Ohio, burst into flames and burned for twenty-five minutes. The burning river quickly became national news. *Time* magazine published an article headlined "The Price of Optimism," complete with a spectacular photo of the river aflame. Randy Newman wrote a song about the famous fire. And decades later, environmental leaders remembered the fire as an emblematic cause of the burgeoning environmental movement. "I will never forget a photograph of flames, fire, shooting right out of the water in downtown Cleveland," President Clinton's EPA administrator Carol Browner said years later. "It was the summer of 1969 and the Cuyahoga River was burning."[1]

But the famous photograph that appeared in *Time* was not of the Cuyahoga River fire of 1969. It was of a far more serious fire in 1952 that burned for three days and caused $1.5 million in damage. In fact, the Cuyahoga had caught fire on at least a dozen occasions since 1868. Most of those earlier fires were much more devastating than the 1969 blaze: A fire on the Cuyahoga in 1912 killed five people. A fire in 1936 burned for five days. The 1969 fire, by contrast, lasted just under thirty minutes, caused only $50,000 in damage, and injured no one.[2] The reason *Time* had to use the photograph of the 1952 fire is that the 1969 fire was out before anyone could snap a picture of it.[3]

For at least a hundred years before 1969, industrial river fires were a normal part of American life. In his scrupulous reconstruction of the era, the environmental law professor Jonathan Adler writes,

> The first reported Cuyahoga River fires were well over a century ago. Indeed, it appears that burning oil and debris in rivers was somewhat common. Due to the volume of oil in the river, the Cuyahoga was "so flammable that if steamboat captains shoveled glowing coals overboard, the water erupted in flames" . . . The Cuyahoga was also not the only site of river fires. A river leading into the Baltimore harbor caught flame on June 8, 1926 . . . The Rouge River in Dearborn, Michigan, "repeatedly caught fire" like the Cuyahoga, and a tugboat on the Schuylkill burned when oil on the river's surface was lit.[4]

It wasn't that nobody had noticed that the river had become a disaster. In 1881, the mayor of Cleveland called the Cuyahoga "an open sewer." The problem was that there wasn't the political will to do much about it. After the Civil War, the city was understandably more concerned with building a new sewer system to prevent more cholera outbreaks than with addressing the occasional river fire.[5]

Like the sad and largely unacknowledged history of the Cuyahoga, smog in Los Angeles and other cities was bad in 1970 but hardly worse than the foul air Americans breathed in earlier eras. All of which begs the question: if modern environmentalism was born in response to the dramatic visual evidence of industrial pollution, why wasn't it born in 1868, 1912, or 1952?

2.

The view of Earth from outer space. The harpooning of whales. The Cuyahoga in flames. Smog in Los Angeles. The clubbing of baby seals. Toxic waste dumps. The hunting of wolves near Yellowstone. The Amazon in flames. Polar bears on melting ice.

Environmental leaders and activists today overwhelmingly believe that these images are the lifeblood of their movement, responsible for motivating the public and policymakers to take action. And so they return again and again to the same idea: if they can just show Americans what is happening to nature, the people will rise up and demand action. Environmentalists believe this to be so because they strongly associate the images of an earlier political moment — the Cuyahoga on fire, the first images of Earth — with the birth and great accomplishments of the modern environmental movement.

In one sense, the dependence upon visual imagery is a kind of nostalgia masquerading as political strategy. And like almost all expressions of nostalgia, it is reductive and simplifies a much more complex picture, ignoring the values and context that defined the moment and obsessively returning to the same partial memories — the exhilaration of seeing images of Earth, the shock and outrage at seeing a river on fire, the imagery painted by Rachel Carson in *Silent Spring* of a world in which the birds had ceased to sing, and the feelings of great accomplishment as millions of Americans poured into the streets demanding action and Congress passed powerful new laws in response.

But in another sense, the overreliance of environmentalists on visual evidence of humans' degradation of nature is a consequence of the environmentalists' interpretive framework; principally, the idea of pollution. Consider that the meaning of the word *pollution* depends on the concept of nature as pure, harmonious, and separate from humans. Pollution is this kind of contamination, or violation, of nature by humans. Similarly, human development is an encroachment upon nature. These are not simply analytical categories but moral ones as well. Nature has been unjustly violated by mankind.

These stories are hardly marginal; they can be found in the most mainstream environmentalist discourses, from Rachel Carson's *Silent Spring* to Jared Diamond's *Collapse* to Al Gore's *An Inconvenient Truth*. Environmentalists are constantly telling nostalgic narratives about how things were better in the past, when humans lived in greater balance with nature. These stories depict humans not as beings as natural as any other but as essentially separate from the world. And while these narratives are easy to recognize, they are difficult to exorcise. They are deeply embedded in the stories environmentalists have long told, the strategies they have become accustomed to using, and the institutions they have built.

But faced with a new set of problems that refuse to reduce themselves so simply to visual explanations of human violations of nature, environmental leaders are at a loss. They complain that the challenge of mobilizing the public to fight global warming is due largely to the fact that global warming is invisible. But the problem is not that global warming is invisible; it's that environmentalists depend too much on the visible.

When all you have is a hammer, the old saying goes, the whole world looks like a nail. Environmental leaders rely on the idea that their political project is to show Americans the ways in which nature is being violated, whether through mailing or beaming those images into Americans' homes, sponsoring nature walks and environmental education programs, or proving through the sciences that human activities are degrading nature.

Environmentalists believe that getting Americans to protect the environment is a simple and rational process: Expose them to the beauty of the natural world. Show them how it is literally being destroyed by human activity. Advocate actions to stop the destruction.

But if getting Americans to see the destruction of nature were enough to galvanize action, why wasn't modern environmentalism born in 1912 or 1952, when fires on the Cuyahoga actually killed people and caused significant damage? If environmental protection is so obvious, natural, and rational a reaction to visible pollution, why didn't the environmental movement begin decades earlier, when pollution was much worse in most American cities? Plainly, something other than a random outbreak of rationality in the 1960s was responsible for motivating the editors of *Time* magazine, Randy Newman, and millions of Americans celebrating Earth Day to demand action to protect the environment.

3.

One of the Sierra Club's most successful coffee-table books, *The Last Redwoods*, published in 1963, explained rising *ecological concern* in the United States in this way:

> [M]an does not live by bread alone. He has needs that are no less real and no less vital — although they are harder to measure in economic terms — than food and water and shelter. He has a thirst for beauty. He often has a hunger for solitude. He craves the companionship of other animal species. He has a deep, atavistic urge for identification with nature. *Witness the extraordinary upsurge of hiking and camping and boating and the overwhelming increase in use of our natural parks.*[6]

It is the kind of statement that likely struck most readers at the time as deeply intuitive: being out in nature is a natural urge. But on further examination, it is an illogical passage. Humans do indeed have a strong urge to be in nature; such an urge might even be atavistic, a long-repressed genetic predisposition. But how can *atavism* explain the extraordinary *upsurge* of Americans hiking, camping, boating, and otherwise enjoying nonhuman nature during the postwar era?

Clearly, it cannot. What explains the postwar increase in recreational boating, camping, and hiking is not atavism — it's *affluence*. The satisfaction of the material needs of food and water and shelter is not an *obstacle to* but rather the *precondition for* the modern appreciation of the nonhuman world.

Between 1945 and the mid-1970s, the standard of living for virtually every American improved consistently and dramatically. By 1970, affluent, comfortable, and secure Americans were strongly interested in quality-of-life concerns, which included things like clean air; clean water; and local, state, and national parks. America's unprecedented postwar prosperity created rising expectations for greater personal fulfillment and a sense that greater material wealth alone could not provide it. Social scientists often label these quality-of-life concerns postmaterial, because they emerge only after individuals and societies have met their basic material needs.

Throughout this book we will distinguish between material and postmaterial needs and values, and so it is important to be clear about what we mean by these terms. While many Americans today could be described as materialistic, that does not mean that they have not yet met their material needs as defined by social scientists. Meeting one's material needs entails meeting one's basic survival needs for food, shelter, and physical security. By this standard, virtually every American today — in contrast to roughly one-third of all Americans in the 1930s — has more than met his or her material needs and is a postmaterialist.[7]

Once we meet our material needs, we all experience a variety of postmaterial needs that are no less strongly felt than our material needs for sustenance and security. This was Abraham Maslow's central insight — that higher needs such as personal freedom, meaningful work, and self-creation do not emerge strongly among those who have barely enough to eat or nowhere safe to live.[8] And while some social scientists and psychologists quibble with elements of Maslow's hierarchy,[9] the underlying concept — that there exists a universal hierarchy of human needs — remains to this day the consensus among social scientists.[10]

Postmaterialists still have material needs. They just spend less time worrying about them than they do their postmaterial needs. Postmaterialists are more likely to worry about eating too much than eating too little. When they worry about where to live, they worry about the quality of life they'll have and not, for the most part, about whether they'll be secure from the elements. The shift from a materialist orientation to a postmaterialist one reflects a shifting of priorities from survival to fulfillment.[11]

The ways in which we satisfy postmaterial needs, and our success in doing so, depend on our beliefs, identities, sociocultural positions, traditions, cultures, values, and moment in time. Throughout this book we will seek to understand differing values and ways of satisfying psychological needs without losing sight of the reality that human needs do not emerge randomly but rather progress in a predictable fashion.

We will also distinguish between lower and higher postmaterial needs, starting with *outer-directed* needs for status, belonging, equality, and freedom and moving to *inner-directed* needs for purpose, mastery, meaning, creativity, and self-creation. Like the progression from material needs to postmaterial needs, higher, inner-directed needs do not appear until people first meet their prior outer-directed needs.[12] One way to think about the rise up Maslow's hierarchy is as the progressive realization of new levels of human freedom: freedom from hunger and deprivation; freedom from violence; freedom to love and belong; and freedom to realize one's potential for creativity.

There are important implications of all this for understanding the birth of environmentalism. Environmentalist values, such as the strong desire to protect ecosystems, largely spring from higher-order, postmaterial, and inner-directed needs. And indeed, around the world there is a very strong association between prosperity and environmental values.[13]

The author of *The Last Redwoods*, which was written at a time when Maslow's ideas were in the air, understood that the appreciation of the outdoors is a human need; what he overlooked was the condition in which it emerged — a telling and ironic oversight, given the demand from environmentalists at the time that we pay attention to the ecosystems in which we live.

The connection between affluence and the birth of environmentalism goes a long way toward explaining why environmentalism in the United States emerged in the 1960s and not in the 1930s. It also explains why ecological concern remains far weaker in Brazil, India, and China than in the United States, Japan, and Europe. And it explains why, when environmentalism does emerge in developing countries, such as Brazil, it does so in Rio de Janeiro's most affluent neighborhoods, where people have met their basic material needs, and not in its slums, where people live in fear of hunger and violence.

4.

The watershed moments that led to the environmental policy victories of the 1970s originated not in 1968, 1969, or 1970 but rather in 1933, 1945, 1960, and 1964. These dates mark, respectively, the inauguration of Franklin Roosevelt and the start of the New Deal; the end of the United States' successful prosecution of World War II; the election of John F. Kennedy; and the devastating defeat of Barry Goldwater by President Lyndon Johnson. These particular historical moments and many others forged the extraordinary affluence and economic growth that characterized the postwar era and made possible the liberal political consensus that largely defined both political parties until the mid-1970s.

The period lasting roughly twenty-five years after World War II was characterized by a consensus between the two major parties about the great issues of the day. The mainstreams of both political parties were avowedly anti-Communist and internationalist, moderately integrationist, and accepting of a brand of benign corporate capitalism made possible by the postwar compact between big business and organized labor. The social upheavals of the 1960s and the economic upheavals of the 1970s fractured that consensus, but it remained operative and dominant in national politics and in the major wings of both political parties through the early 1970s.

Contrary to popular stereotypes, environmentalism was never particularly countercultural but was a product of the liberal opinions of the era. America's most important environmental bills were passed by the liberal establishment and signed into law by Republican presidents. Earth Day itself was the invention not of idealistic young people concerned about the fate of the planet but rather of powerful Washington, D.C., liberal insiders. Democratic Wisconsin senator Gaylord Nelson came up with the idea after reading about campus teach-ins against the war in Vietnam. He asked for and received office space from the liberal good government group Common Cause; not long after, the Ford Foundation provided the NRDC with its founding grant. It was a powerful liberal senator, not student radicals, who came up with the idea for Earth Day. And it was Ivy League universities, not rural communes, that produced today's environmental leaders.

Consider President Nixon's extraordinary 1970 State of the Union speech, where his discussion of happiness, the environment, and quality-of-life concerns took up the final third of the speech.

In the next 10 years we shall increase our wealth by 50 percent. The profound question is: Does this mean we will be 50 percent richer in a real sense, 50 percent better off, 50 percent happier? Or does it mean that in the year 1980 the President standing in this place will look back on a decade in which 70 percent of our people lived in metropolitan areas choked by traffic, suffocated by smog, poisoned by water, deafened by noise, and terrorized by crime? . . . The great question of the seventies is, shall we surren- der to our surroundings, or shall we make our peace with nature and begin to make reparations for the damage we have done to our air, to our land, and to our water?

It was a speech that wouldn't have made sense during the material privations of the Great Depression and World War II. It was a speech that would be almost unimaginable coming from a politician of either major party today. Nixon was neither a liberal nor an environmentalist, but he lived and governed at a time when the social values and politics on both sides of the partisan divide were profoundly different. Those values were the reflection of a particular political moment. Despite the divisions of the Vietnam War, most Americans had within their lifetimes witnessed the striking economic and social transformations of the nation from Depression-era scarcity to postwar abundance.

It is no accident that the great accomplishments of modern environmentalism, along with the great accomplishments of postwar liberalism, occurred at the moment of twentieth-century America's greatest prosperity, nor is it a surprise that they occurred in the shadow of the oil shocks of the mid- and late 1970s that brought the egalitarian economic expansion of the postwar era to an end and the U.S. economy to its knees. Those economic shock waves, along with the global economic forces that transformed the American economy in the 1980s and 1990s and the rising conservative tide that dismantled much of the social safety net that the New Deal and the Great Society had built, brought about a new era of economic organization. With this new era came a new social values environment and a new political realignment that liberals and environmentalists alike have largely failed to address.

5.

Faced with two decades of evidence that their agenda has failed to be particularly powerful politically, environmentalists fall back on superficial polling showing that strong majorities of Americans support action on environmental issues. But when independent researchers delve beneath the surface of those attitudes, they find that while the public supports the environmental agenda, they don't support it very strongly. In 2005, Duke University's Nicholas Institute for Environmental Policy Solutions hired a Republican pollster and a Democratic pollster to look more closely at the relationship between environmental attitudes and political behavior. Their research found that 79 percent of Americans interviewed favored "stronger national standards to protect our land, air and water" — the same easy-to-agree-with statement that tends to reassure environmental leaders.

But the Nicholas Institute's pollsters also found that only 22 percent of voters said environmental issues played a major role in how they voted. Even among those voters who were upset that more wasn't being done to address their environmental concerns, just 16 percent reported feeling angry about it — 70 percent reported the milder feelings of disappointment and concern. Perhaps the most important finding of all was this one: even self-identified environmentalists prioritize other issues — gay marriage, abortion, and illegal immigration — ahead of the environment.[14]

When the pollsters asked voters to rank issues in terms of their importance, the environment almost always came in last. In the Nicholas Institute's survey, pollsters asked, "What is the most important issue to you personally?" The environment came in dead last (10 percent of voters), after economy/jobs (34 percent), health care (25 percent), Iraq (22 percent), Social Security (21 percent), education (20 percent), terrorism (20 percent), moral values (15 percent), and taxes (12 percent).[15]

In a June 2006 survey, researchers with the Pew Research Center for the People and the Press asked people what issues they considered very important. The environment came in twelfth (51 percent) and global warming sixteenth (43 percent) out of sixteen named issues. The percentage of Americans saying that the economy was important to them came in at 79 percent, and jobs came in at 68 percent. By January 2007, after the alleged tipping point in opinion, global warming had actually declined in importance for voters.[16]

It is worth noting that the environment and global warming get that minimal level of support only when they are named in the poll as options to choose from. In 2003 we conducted a poll of voters in Pennsylvania — one of the handful of battleground states that determine who is elected president — and asked the open-ended question "What is the most important concern facing your community?" Just 2 percent named the environment. Not a single person named global warming.[17]

In poll after poll, voters name concerns like jobs and the economy as their top priorities. Yet environmental foundations and organizations haven't advanced a strategy for action on the environment to be centrally justified as a way to create jobs and stimulate growth. The simple reason for this is that the categories of jobs and growth exist outside of the environmentalist politics of limits.

Given the environmentalists' unwillingness to speak to what really matters to most Americans, it should come as little surprise that the perception of environmentalists has deteriorated dramatically since the early 1990s. The Environics social-values survey, which has been conducted in the United States and Europe since 1992, has found that the number of Americans who agree with the statement "To preserve people's jobs in this country, we must accept higher levels of pollution in the future" increased from 17 percent in 1996 to 27 percent in 2004. The news is even worse for environmentalists themselves. The number of Americans who agreed that "Most of the people actively involved in environmental groups are extremists, not reasonable people" leaped from 32 percent in 1996 to 43 percent in 2004.[18]

6.

Having grown up in the corridors of power, environmental leaders from the generation of '68 were never forced to grapple with the kinds of existential and philosophical questions pondered by those currently holding the short end of the historical stick. "Liberals are less conscious of public philosophy," the conservative columnist David Brooks observed, "because modern liberalism was formed in government, not away from it." Conservative political impotence in the early seventies was the mother of antienvironmental invention. Out of power, young conservatives got clear that social policy and government action did not result from identifying the best, most efficacious, or most rational policies but rather were an expression of social values, public morality, and political power. Wrote Brooks:

> When modern conservatism became aware of itself, conservatives were so far out of power it wasn't even worth thinking about policy prescriptions ... Conservatives fell into the habit of being acutely conscious of their intellectual forebears and had big debates about public philosophy. That turned out to be important: nobody joins a movement because of admiration for its entitlement reform plan. People join up because they think that movement's views about human nature and society are true.[19]

Most of today's national environmental leaders and philanthropists joined the movement at a time when liberals and environmentalists controlled the government and the levers of power. They thus treat legislation as simply an effort to fix problems — not as a way to advance a particular morality and worldview.

Befitting members of a movement who came of age at the height of the liberal postwar consensus, most environmentalists are largely unconscious of the social and historical forces that shaped their values and identities. They are, in this way, rationalistic. Like other liberals, environmentalists tend to believe that they have arrived at their values and worldview through a rational and considered process. They rarely struggle with the myriad nonrational and unconscious motivations for telling the stories they tell and believing what they believe.

In the stories they tell about their movement's birth and in the politics they make for its future, environmental leaders steadfastly ignore the central role that evolving values play in shaping society and politics. They see the postwar increase in outdoor activities as evidence of something ancient rather than something new. They view their movement as a reaction to new, visible pollution problems, such as the Cuyahoga River fire of 1969. They treat their policy victories, such as the Clean Water Act, as rational responses to new pollution problems of industrialization, and they overlook the ways in which the act was a manifestation of changing social values made possible by industrialization. And they credit the nonhuman sciences for having revealed to the public environmental degradation and ignore the social and political environment that supported the public's faith in science.

7.

Just as prosperity tends to bring out the best of human nature, poverty and collapse tend to bring out the worst. Not only are authoritarian values strongest in situations where our basic material and security needs aren't being met, they also become stronger in societies experiencing economic downturns. Economic collapse in Europe after World War I, in Yugoslavia after the fall of communism, and in Rwanda in the early 1990s triggered an authoritarian reflex that fed the growth of fascism and violence. The populations in those countries, feeling profoundly insecure at the physiological, psychological, and cultural levels, embraced authoritarianism and other lower-order materialist values.[20] This is also what occurred in Iraq after the U.S. invasion.[21]

This shift away from fulfillment and toward survival values appears to be occurring in the United States, albeit far more gradually than in places like the former Communist-bloc countries. Survival values, including fatalism, ecological fatalism, sexism, everyday rage, and the acceptance of violence, are on the rise in the United States.[22] The reasons for America's gradual move away from fulfillment and toward survival values are complex. Part of it appears to be driven by increasing economic insecurity. This insecurity has several likely causes: the globalization of the economy; the absence of a new social contract for things like health care, child care, and retirement appropriate for our postindustrial age; and status competitions driven by rising social inequality.[23]

Conservatives tend to believe that all Americans are getting richer, while liberals tend to believe that the rich are getting richer and the poor are getting poorer. In our discussion of security in chapter 7 we argue that what is happening is a little bit of both: homeownership and purchasing power have indeed been rising, but so have household and consumer debt and the amount of time Americans spend working. While cuts to the social safety net have not pushed millions of people onto the street, they have fed social insecurity and increased competition with the Joneses.

It is not just environmentalists who misunderstand the prosperity-fulfillment connection. In private conversations, meetings, and discussions, we often hear progressives lament public apathy and cynicism and make statements such as "Things are going to have to get a lot worse before they get better." We emphatically disagree. In our view, *things have to get better before they can get better.* Immiseration theory — the view that increasing suffering leads to progressive social change — has been repeatedly discredited by history.[24]

Progressive social reforms, from the Civil Rights Act to the Clean Water Act, tend to occur during times of prosperity and rising expectations — not immiseration and declining expectations. Both the environmental movement and the civil rights movement emerged as a consequence of rising prosperity. It was the middle-class, young, and educated black Americans who were on the forefront of the civil rights movement. Poor blacks were active, but the movement was overwhelmingly led by educated, middle-class intellectuals and community leaders (preachers prominent among them). This was also the case with the white supporters of the civil rights movement, who tended to be more highly educated and more affluent than the general American population. In short, the civil rights movement no more emerged because African Americans were suddenly denied their freedom than the environmental movement emerged because America suddenly started polluting.

8.

While social scientists have demonstrated that material prosperity is a prerequisite for ecological concern, environmentalists point to figures like the Brazilian rubber tapper and union leader Chico Mendes, the Nobel Prize–winning Kenyan Wangari Maathai, and environmental justice activists in the United States as proof that environmentalism is as much a politics of the poor as the rich, largely ignoring the far higher priority concerns that Brazilians, Africans, and poor communities in the United States have. As a result, it has been conservative scholars, like Jonathan Adler, whose Cuyahoga River history we cited above, and not environmental or liberal scholars, who have paid the most attention to the role of material prosperity in increasing ecological concern.

Environmental leaders and philanthropists, as we shall see, treat prosperity as either a distraction from their primary focus of solving the problems of pollution and bad development or as an ancillary concern to be jury-rigged on the existing environmental framework of limiting human intrusions on nature. But improving living standards for humans can never become a fundamental aspect of environmentalism — only a weakly grafted appendage. That's because environmental thinkers and leaders understand material prosperity as the *cause of but not the solution to* pollution and degradation.

In the case of the Amazon and global warming, environmental

leaders treat modernization as an obstacle to dealing with ecological crises. They see global warming and the destruction of the Amazon as crises of *too much* rather than *too little* progress. When environmentalists do embrace economic development, it is almost always as one more tool for protecting nonhuman natures. Conservationists have tended to care about the debts incurred by dictatorships in Latin America and Africa only to the extent that they can be swapped for land the conservationists want to protect. They care about economic development *only* to the extent that it is sustainable and occurring inside those areas on earth they call nature. When they think of economic development at all, it is almost always in the service of an immediate exchange or, at best, a conditional acceptance.

Environmental leaders have long seen the newfound prosperity of countries like China and India as a threat to ecological stability, not as something that may bring new values and new politics that can be used to address global warming and other ecological crises. If they see opportunity at all, they see it in the same technical fixes they peddle to the rest of the developed world: fuel-efficient cars, fluorescent light bulbs, and better appliance standards.

In failing to direct their gaze at the awesome human potential in China and India, environmental leaders are constantly missing new possibilities for creating a politics that speaks to the economic aspirations that citizens of developed countries take as their birthright.

In denying the central role that prosperity has played in giving rise to their politics and in failing to make material prosperity for all a central part of their politics, environmentalists have to defend themselves against the accusation that they are elitists who are more concerned with protecting beautiful places than in improving the lives of the billions of people around the world who still live in abject poverty. When environmental leaders respond that the poor suffer the most from pollution and benefit the most from environmentalism, they continue to miss the point: however bad the pollution and loss of nonhuman natures may be, hunger and insecurity are almost always more strongly felt.

All of this has contributed to the caricature of environmentalists as overly sensitive to the fate of snail darters and spotted owls and insensitive to the plight of people. The truth is that those environmentalists who take antimaterialist positions on everything from global warming to local development do so not because they are hardhearted but because they are narrow-minded. They are so caught up in explaining ecological crisis as a consequence of human intrusions upon nature that the only solutions they can imagine have to do with reducing the frequency and severity of these intrusions.

9.

The 2006 election and the return of Democrats to power in Congress have given many environmentalists and liberals comfort. But we see no evidence that Americans are any more secure socially and economically than they were before the election. And as long as so many Americans are fearful, pessimistic, and insecure, America's social values ecology is unlikely to support the kinds of changes required to deal with global warming.

Given the undeniable connection between ecological concern and the sociocultural context that makes it possible, one would think that environmental strategists and funders would be focused on caring for the social and political ecology that are necessary preconditions for an effective and powerful environmental politics. But instead, they have defined environmentalism as an issue category that concerns itself with representing nature's interests and defending them from human intrusion.

Having told itself a reductive story about the past, the environmental movement unsurprisingly engages in a reductive politics in the present. Convinced that their past victories resulted from the public's seeing pollution, environmental leaders constantly search for new ways to show the public evidence of environmental degradation. Understanding the birth of the environmental movement as a reaction to industrialization rather than a product of it, environmentalists imagine solutions that seek to constrain, rather than unleash, human activity and economic growth. And having constructed personal and institutional identities based on this birth story, environmentalists find themselves unable to move to alternative strategies or stories about how we need to change the world.

The present moment offers us an opportunity. With control of Congress and probably better than average odds at recapturing the presidency in 2008, progressives of all stripes have an opportunity to retool our politics for a new century. But doing so will demand that we avoid imagining that the good old days of sixties-era consensus liberalism are back and that the last three decades have been a bad dream from which we have finally awoken.

Creating a new progressive political consensus requires paying close attention to social values, how they are evolving, and how we might create a new social contract for postmaterial America that can provide enough security and prosperity to support a new, more ecological era.

2

THE FOREST
FOR THE TREES

On JULY 24, 1993, millions of people worldwide awoke to newspaper photographs of dead children wearing little more than rags lying in copious amounts of blood in the shadow of Nossa Senhora da Candelária, one of Rio de Janeiro's most cherished cathedrals.

The children, some of whom had been on the streets since they were three years old, dozed each night in front of the cathedral located in the heart of downtown Rio. Whether for lack of food or fear of violence, they could no longer sleep at home. Huddled together for warmth on flattened cardboard boxes and soiled newspapers, the children thought they were safe under the protective arches of the church.

For Officer Marcos Emmanuel and seven other off-duty cops, the children who took refuge each night at the Candelária were nothing more than filthy street vermin. It was obvious to him and everyone else that they were behind the street crime in the area — the indecorous begging, the stealing from merchants, the stickups of tourists. When word got out that a pack of them had pummeled a local patrol car with rocks, trying to shatter the windows and the resolve of the men inside, Emmanuel and the other officers decided to take dramatic action.

The children were asleep when the cops pulled up. Emmanuel

and another officer grabbed three of the older kids who they suspected were the gang's ringleaders, shoved them into the car, and shot
each of them in the head three times. They dumped the bodies on a
lawn across from the Museum of Modern Art a short distance away
and sped back to the Candelária. On their return they found that the
other officers had nearly finished shooting forty children. By the end
of the evening, eight of the children were dead.

The next day's news coverage felt, for most Brazilians, like a stiff
brush over an infected wound. The occasional extrajudicial killing of
delinquents was one thing; doing it under the protective gaze of
Nossa Senhora was quite another. Even those Brazilians who had
been ranting at the nearby street-corner soda counters about the an-
noying *molecas* couldn't look at the newspaper photos and not feel
disgusted by the whole thing: the kids, the cops, the country. There
was no shortage of Brazilians who called radio stations the next day
to say that they thought it was about time somebody cleaned up the
street filth. But for most, the spectacle summed up what Brazilians
hate most about their country.[1]

Thanks to the audacity and symbolic setting of the crime, and
the fact that July 23, 1993, was a slow news day internationally, demands that the government do something poured in from abroad.
Brazil's social movement of street children and their allies held dem-
onstrations throughout the country. Politicians promised to root out
the evildoers, send them to prison, and reform the system. Strongly
worded editorials were written. International conferences were held.
A few low-level officers went to prison. A few midlevel officers were
let go or transferred. In short, nothing of consequence was done.[2]

Around five children continue to be murdered every day in
Brazil. Their deaths are for the most part quiet. Of the sixty-two children who survived the Candelária massacre, thirty-nine are now dead
— likely more, by the time you read this — from drugs, guns, or
police.[3] What was unusual about Candelária was that it got any attention whatsoever.[4] When a record thirty civilians were randomly

gunned down on March 31, 2005, by off-duty cops seeking blanket revenge for fallen comrades, hardly anybody outside of Brazil heard about it. Two days later, massacre-fatigued Brazilians had the death of Pope John Paul II to grieve.[5]

The colossus of the south today continues to offer up sensational killings that occasionally attract international attention, and the street children's movement is a firmly established and respectable interest group. But more often than not, everyday acts of terrorism, from the murder of small children to the police shootings of bystanders, are, like the burning of the Amazon, little more than background noise in our cacophonous modern world.[6]

1.

The killing of street children is but one manifestation of the violence that keeps half of all Brazilians living in poverty. Most street children are not orphans. They have parents and even homes. They just can't, for lack of food or the certainty of abuse, live in them.

The everyday violence in Rio de Janeiro is a mirror of the destruction in the Amazon. Both are manifestations of extreme poverty and inequality. Brazil has the world's highest death toll from guns — about 36,000 per year.[7] Many Brazilians carry guns for self-defense because the government cannot guarantee personal safety. And recently, drug gangs operating from within Brazil's prisons have started unleashing terrorist attacks against civilian targets, including churches, in order to win better conditions from the government. More than 180 people were left dead from attacks in May 2006. Because the gangs have better weaponry than the police — such as grenades and mortars — the federal government had to send ten thousand army troops into São Paulo.[8]

Seeking protection as much from the police as from criminals, many of the poorest people in Brazil reluctantly welcomed heavily armed drug traffickers into their communities in the 1990s. "Drug trafficking has a good and a bad side to it," one Rio *favela* (slum) resident told a group of documentary filmmakers in 2000.

> Before there were dealers, when the police entered the *favela* they would break down doors. So, when these guns entered the community from the drugs, they forced the police to behave more cautiously. Now, the negative side, the cruelty of the guns, is that when they come to collect debts from people in the community, they don't care if that person is a minor or not. They will kill and cut up and quarter that person to display as an example for everyone to see.[9]

And yet Brazil is a fantastically wealthy country. Larger in size than the continental United States, it boasts sophisticated manufacturing, mining, biotechnology, and agricultural industries. It is a major exporter of coffee, soybeans, iron ore, orange juice, steel, and even high-tech airplanes, which it manufactures for companies like the U.S. airline JetBlue.

Few are more articulate about the nexus of lawlessness, corruption, and violence than Brazil's cops. The former chief of police of Rio de Janeiro explained the situation this way:

> The police are corrupt. The institution was designed to be violent and corrupt. Why do I say this? Because it was created to protect the state and the elite. I practice law enforcement to protect the *status quo*. There's no beating around the bush about this. It keeps the *favelas* under control. How do you keep two million underprivileged people [in Rio] who make 112 *reais* [less than $50] per month under control? With repression, of course.[10]

Brazil's richest 10 percent of the population has an annual income nineteen times higher than that of the poorest 40 percent. For poverty to be managed in a society as wealthy as Brazil, what's required is, in the words of the police chief, an institution "designed to be violent and corrupt."[11] It is through this prism of violence, poverty, and inequality that the destruction of the Amazon must be understood.

2.

Land in Brazil is like oil in Nigeria: as much a curse as a blessing. Its control by the few denies prosperity to the many. The most affluent 1 percent of Brazilian landlords own 45 percent of the land, some of it inherited, some of it bought, and much of it stolen. Today, roughly 4.5 million peasants are without land; many millions more have left the countryside to live in the 100,000-plus-person slums, or *favelas*, of the country's largest cities.[12]

Since its founding by the Portuguese crown, Brazil has long hoped to colonize its seemingly limitless interior. The country's 1891 constitution stipulated the construction of a new capital in the center of the country that was meant to replace Rio de Janeiro. The idea was that the new city would be a launching pad for development northward into the Amazon forest, an area eleven times the size of France. By 1893 a site was selected, and in 1955 President Juscelino Kubitschek announced plans to create a new capital in the parched, empty center of Brazil, calling for "fifty years' progress in five."[13] Kubitschek wanted Brasília, the new capital, to catapult Brazil into the modern world.

But Brasília and the expansion into the Amazon came at a terrible price. Because Brazil had little money to pay for the new capital, Kubitschek simply printed more money. This triggered Brazil's long struggle with inflation, which rose from 9 percent in 1950 to 58 percent in 1965 to 235 percent in 1985 to an astonishing 1,783 percent in 1989.

In 1964, taking advantage of the country's economic and political chaos, Brazil's generals staged a coup d'état with the support of the U.S. government.[14] Upon seizing power, they were quickly confronted with the same confounding thicket of impossible demands, intractable conflicts, and irremediable problems that every Brazilian leader, militarily imposed or democratically elected, has faced since. Rampant inflation and crushing debt forced Brazil's leaders to choose between satisfying the demands of foreign investors — and thus cutting public investments in education, health care, and emergent industries — or making the public investments and risk being cut off from the foreign capital and markets they needed.

Looking for a way out of the nation's many conflicts and contradictions, Brazil's generals, like generations of Brazilian leaders before and after, placed their faith in the one seemingly inexhaustible resource that has been an endless source of fascination and fantasy for five hundred years: the Amazon forest.

From a strictly economic point of view, Brazil would have been better off if it had left the Amazon alone and focused its agricultural and economic development elsewhere. But massively unequal land ownership — a residue of the colonial land system set up by the Portuguese crown — concentrated national wealth in impractically large, idle estates, some the size of small European countries, resulting in almost ritualized conflicts between peasants and Brazil's powerful landlords. Government leaders who attempted meaningful land reform faced the wrath of landowners, who, through corruption and other forms of political influence, had long been a part of the state itself.

Development of the Amazon seemed to offer an alternative to the impossible conflicts facing the nation, and Brazil's generals jumped at the prospect, believing that in doing so they would entice foreign investment, earn foreign capital, grow the economy, and obviate the need for land reform all at the same time.

In 1966, Brazil's dictator president, General Castello Branco, announced that "Amazonian occupation will proceed as though we are waging a strategically conducted war," branding the colonization efforts as a national project of unification — a march toward *ordem e progresso*.[15] The effort included massive tax breaks for investors, huge development projects such as the paving of the Belém to Brasília highway and the construction of massive new dams, and an Amazonian version of the Homestead Act to encourage the colonization of the Amazon by settlers. Aping the old Israeli slogan, Brazil's generals declared the Amazon to be "a land without people for a people without land."

The result was a disaster by everyone's estimation, including the generals'. Brazil's elite sensed a land rush and began buying up land in the region. Hopeful homesteaders arrived in the Amazon only to discover that people had already made claims to their land, often fraudulently. Because the easiest and in some cases the only way of proving ownership was to demonstrate use, the first thing farmers, large and small alike, did was log and burn the rain forest. Given the utter lack of law enforcement in the area, landlords blithely tortured and killed peasants who got in their way. Brazilian peasants who cleared land for cattle and farming discovered after a few years that slash-and-burn agriculture couldn't sustain them, and so millions fled to the overcrowded *favelas* of nearby cities.[16]

3.

Brazil's inflation, its debt, and its inequitable land distribution would become the macrodrivers of hunger, misery, and deforestation for a half century to come. In response to the widely publicized failures of colonization, Brazil's military rulers shifted their emphasis to massive infrastructure development — roads, dams, and industrial agriculture — financed with loans from private lenders, the World Bank, and the Inter-American Development Bank.

After the first OPEC oil shock, in 1973, the military took the nation's debt to a new level when the dictatorship borrowed billions from foreign lenders at low but also *variable* interest rates in order to cover the abrupt increase in fuel costs. The generals' borrowing accelerated after the second oil shock, in 1979, when oil accounted for 43 percent of Brazil's imports.[17]

The moment of reckoning arrived in 1981, when the U.S. Federal Reserve raised the prime rate. By 1982, having exhausted nearly all of its hard currency for foreign exchange, Brazil defaulted on its loan payments to its private creditors.[18] The consequences were harsh, both for the Brazilian people and their national treasure, the Amazon forest. From 1982 on, Brazil spent much of its foreign exchange reserves servicing the foreign debt. Brazil repeatedly rolled over its debt, accepting the burden of paying more interest in the future to make smaller payments in the present.[19]

As of 2007, Brazil was one of the world's largest debtors, owing an astonishing $511 billion[20] — an amount it has paid off several times over even as the principal has increased.[21] One disastrous result of Brazil's enormous debt is a skyrocketing rate of deforestation. By 1990, the year that Brazil discontinued direct subsidies for cattle ranching, roughly 587,000 square kilometers (317,000 square miles) of rain forest had been destroyed.[22] And the number of non-Indian Brazilians living in the Amazon increased from ten million in 1970 to twenty-one million in 2000.[23]

The Amazon remains the top destination of poor Brazilian migrants seeking a better life. Many come from the impoverished northeastern states of Maranhão and Piauí, often escaping the violence of landlords in already colonized states. One of Brazil's leading Amazon reporters, *Veja* magazine's Leonardo Coutinho, told us, "Impoverished and punished by violence, they come to the Amazon to serve as semi-enslaved workers clearing the forest. In some cases they become hired guns themselves."[24]

The criminal mafias who oversee illegal logging and deforestation — and the murders of peasant leaders — often also traffic in drugs and guns. Even when Brazil's financially strapped and corrupt environmental enforcement agencies crack down on illegal deforestation, they find themselves badly outmatched by land speculators and cattle ranchers, who use sophisticated satellite equipment and deforest in stages in order to evade law enforcement.[25] "Even when the government is able to spot illegal deforestation using satellite images," Coutinho points out, "it usually takes a week for agents to arrive in the field. By that point, the loggers have taken what they wanted and moved on. The state simply doesn't function out here. The law of the Amazon is made by the bullet."[26]

One of Brazil's most frustrating paradoxes is that it has some of the most progressive environmental and human rights laws on the books but is mostly unable to enforce them. It is illegal to export hardwoods from the Amazon, and yet every year Brazil sells nearly $400 million worth of them to foreign buyers.[27] Slavery was officially banned in 1888, but today somewhere between twenty-five and fifty thousand souls toil away as slaves in gruesome conditions, logging mahogany and other hardwoods for sale on global markets.[28] For the most part, this is a consequence of Brazil's economic reality: when pushed to choose between servicing its foreign debt and investing in law enforcement, it chooses to service its debt.

The Brazilian government simply doesn't have the resources to eliminate slavery or govern the *favelas* in Rio patrolled by drug traffickers, much less properly enforce land and environmental laws in the Amazon. Reassured by the virtual absence of state power in the Amazon, Brazil's powerful agricultural, logging, mining, and other interests hardly bother opposing environmental laws and regulations in Brasília.[29]

Everyone acknowledges that there are macroeconomic drivers of deforestation, from the government's infrastructure projects to cattle and soy exports. And yet Brazil's debt is the elephant in the room that few environmentalists discuss — much less prioritize — even as the global movement to "make poverty history" through debt relief gains momentum.

It has long been obvious that what drives the murder of street children is what drives the murder of peasant leaders — and that what drives the growth of *favelas* in Rio is what drives the decimation of the Amazon. Why then do Brazilian and non-Brazilian environmentalists alike make the violence and poverty in the Amazon part of their politics but not the poverty and violence that afflict the rest of Brazil?

4.

In the late 1980s the world learned the saga of a theretofore obscure Brazilian rubber tapper named Chico Mendes. Every drama needs its hero, and Mendes made an excellent one. His story could trigger a tear among North Americans and Europeans who knew nothing of Brazil or the Amazon. He was the ultimate underdog and martyr. From his death was born the first major global effort to protect the Amazon, one founded on the idea that rubber tappers and Indians, who had been enemies, would together lead a movement to lift the Amazon's people out of poverty and protect the forest.

Around the world, a barrage of hopeful books and articles were published touting the romantic notion that the little guys of the forest were agents of history, authentically representing nature's true interests against corrupt governments, violent landlords, and rapacious corporations. Even writers critical of how Europeans projected their own romantic notions of the "noble savage" on Indians ended up reading their own fantasies into Mendes's coalition. The following is a typical example from the period:

> They called for popular control over the means of production and distribution of forest commodities, along with the provision of financial credits to producers rather than to middlemen. They also called for justice and legal protection of their rights to land and life. These are the concrete elements of a socialist ecology — the *only* strategy that can save the Amazon and its inhabitants.[30]

By the time of the 1992 United Nations environment conference in Rio de Janeiro, everyone from CEOs to ranchers to heads of state had adopted the language of sustainable development, leading environmentalists to believe that the promises being made would somehow progress to a new mode of economic development in the Amazon. It was a belief reinforced by several hundred million dollars invested by wealthy countries in conservation and sustainable development pilot projects, with $200 million invested by computer-chip mogul Gordon Moore alone.

But in romanticizing Chico Mendes and the movement he helped create, environmentalists largely missed the most important lessons from his life and death. What few people outside of Brazil realize is that Mendes was a labor and community organizer, not an environmentalist. Passionate, intelligent, and largely self-educated, Mendes adopted the discourse of environmentalism for tactical reasons. He realized that doing so offered the opportunity to draw attention, support, and resources to his objectives. Just before he was gunned down, Mendes told an interviewer, "Until recently, the word 'environment' didn't even exist in our vocabulary."[31] *New York Times* reporter Andrew Revkin observed, "His demands were for schools and jobs and health care, hardly a green agenda. This overlap with environmental preservation brought the union man to the attention of conservationists who shared his goal of preserving the rain forest, but for far different reasons."[32]

Revkin quotes Mendes yelling back at a TV documentary that depicted the rubber tapper as an environmentalist. "I'm not protecting the forest because I'm worried that in twenty years the world will be affected," Mendes said. "I'm worried about it because there are thousands of people living here who depend on the forest — and their lives are in danger every day."[33] The successor to Mendes as the head of the rubber tappers union, Osmarino Amâncio Rodrigues, so feared that environmentalism was distracting attention away from the need for land reform that he denounced "empty environmentalism" in a provocative 1990 essay titled "The Second Assassination of Chico Mendes."[34]

In contrast to conservationists who define the challenge as "protecting nature," *Veja* magazine reporter Coutinho, like Mendes before him, insists that the key to protecting the Amazon is achieving a high enough level of prosperity so that Brazilians can care for their forest. "It's inconceivable to think that Brazilians who work with medieval farming technologies and businessmen who operate with a 300-year-old mentality will ever be able to care for this place," he told us. "The Amazon will only be saved when people who live here reach a minimum level of development."[35]

It is unreasonable to expect individuals whose basic material needs haven't been met to care strongly about the nonhuman world. Likewise, until the populations of China and Brazil have achieved a minimal level of economic development and security, it is unreasonable to expect those countries to sacrifice economic development for the purposes of reducing pollution and protecting nonhuman ecosystems. The new mentality Coutinho is hoping for includes not just ecological concern but also respect for workers, long-term planning, and more efficient use of resources — postmaterialist values that tend to emerge together and support a kind of development that is good for humans and nonhumans alike.

Unfortunately, much of the advocacy for sustainable development ignores the fact that ecological concern is a postmaterialist value that becomes widespread and strongly felt — and thus politically actionable — only in postscarcity societies. In the case of Brazil, sustainable development has been defined strictly as pilot projects *inside* of the Amazon — even though such projects often end up attracting migrants into the forest and accelerating deforestation. Economic growth in big cities outside of the Amazon, such as São Paulo, does more to take pressure off the Amazon than development within it, but because São Paulo is outside of those areas understood as natural, it has no obvious place in environmental politics.

5.

Part of the reason environmentalists continue to believe their small-bore approach to dealing with massive global problems will work is that they think it already does. There are few more insidious myths that conservationists tell themselves than that they have already saved 40 percent of the Brazilian Amazon from destruction.[36]

Technically, the Brazilian government mandates that a large percentage of all privately held forestlands be left standing.[37] But with such lax enforcement, most producers deforest 100 percent of their land with no consequences. Because the easiest way to prove ownership of the land is to deforest it, conservation is mostly against the self-interest of property owners.

Pervasive lawlessness is an essential aspect of Brazilian reality, and yet conservationists cling to the hope that the government will, somehow, for some reason, find the resources and decide to invest them in enforcing national and state laws protecting the forest from deforestation. They believe this despite a history of nonenforcement and the low priority the government gives to environmental issues. To put it another way, the 40 percent of the Amazon in parks, reserves, and indigenous areas is about as safe from destruction as a street child in Rio is from being shot.

Given the Brazilian government's track record, its current economic model, and its financial commitments, it is unlikely that they plan to protect anything close to the 70 percent of the Amazon that scientists say must be protected to prevent the forest from drying out and spiraling into uncontrolled burning. The Brazilian government doesn't say so directly, but there is little reason to believe it would not let the destruction continue past the 30 percent mark — unless of course there was some other way to generate revenue and growth.

But even if the government, the tribes, and the managers of extractive reserves manage over the next several years to keep 40 percent of the forest that has been saved intact, what then? The tighter Brazil's protections on the first 40 percent of the forest are, the greater the pressure will be on the remainder of the forest. In 1987 there was more than enough land available for cattle ranching, farming, and conservation. But today, the pressures for development are only increasing.

"If the vision is just around setting up parks and prohibiting things, it isn't going to work," observed Anthony Anderson, an American scientist who lived in Brazil for many years and recently worked as a consultant to the U.S. Agency for International Development. "Maybe we'll get to 40 percent protection. But without engaging Brazilian civil society, and rising up to a higher place in the national political debate, conservation isn't going to happen."[38]

6.

Despite the international outcry following Mendes's death in 1988, the Amazon's destruction continued faster than ever before. Whereas 21,000 square kilometers (13,500 square miles) were deforested between 1977 and 1988, an astonishing 84,000 square kilometers (52,000 square miles) were deforested from 2002 to 2005.[39] Today, approximately 20 percent of the forest is gone,[40] and the equivalent of eleven football fields of Amazon rain forest is being destroyed every minute, the fastest rate of forest destruction anywhere.[41]

Brazil is an agricultural powerhouse, providing 80 percent of the world's orange juice, 38 percent of its soy, and a third of its sugar, chicken, and coffee. It is the world's number-one beef exporter. In 2004, agriculture represented 31 percent of the country's GDP. Brazil had $27 billion in agricultural exports, giving it a trade surplus of $24 billion — the largest in the world.[42]

As international demand for Brazilian beef and soy has grown, the Brazilian government has encouraged the expansion of soy farming into the Amazon because it provides one of the few reliable sources of hard currency, which the government needs to service its debt.[43] Today, the Brazilian government supports massive infrastructure projects in the Amazon, from dam building, road paving, agriculture, and mining to moving entire rivers, through a program called Avança Brasil — Forward Brazil. The government said it would invest $20 billion in these programs by 2007.[44]

One of the government's top Avança Brasil priorities is to pave the 1,100-mile-long Amazonian highway to more easily transport soybeans from the soy-growing state of Mato Grosso to the Amazon port city of Santarém and from there to the United States, Europe, and Asia, where demand for soy livestock feed has skyrocketed.[45] Experts estimate that paving this one road will lead to the deforestation of eight hundred kilometers of forest alongside the highway.[46] As a sign of how small a priority conservation is to the Brazilian government, consider that the $140 million conservation package the government announced in 2004 is less than *1 percent* of Avança Brasil's total budget.[47]

Scientists estimate that Avança Brasil, depending on how fully it is implemented, could lead to a loss of 42 percent of the forest by 2020.[48] The nonoptimistic scenario of 42 percent loss means that by 2020, "few pristine areas will survive outside the western quarter of the region" and only about 5 percent of the total Amazon would be "free from non-indigenous impacts."[49]

While the environmental movement's microdevelopment projects in the forest arouse little interest outside of conservationist circles, Avança Brasil conjures hopes of one day overcoming poverty. "Nothing threatens the rain forest as much as poverty and ignorance," wrote Avança Brasil's secretary of planning José Paulo Silveira in the journal *Science*, attacking the scenarios predicting the program could wipe out 42 percent of the forest. "Some 20 million people live in the Brazilian Amazon region, most of them very poor. We must offer these people a lifestyle better than hacking and burning. That is what Avança Brasil seeks to do."[50]

Frustrated by the government's tacit support for deforestation, some conservation leaders are hoping to create a market for sustainable soy and beef. They also hope to pressure European and American buyers of Brazilian cattle and soy, such as food giants Unilever and Archer Daniels Midland, to purchase sustainably produced beef and soy only, or to force their suppliers to respect Brazil's environmental laws, especially the requirement that 80 percent of privately held lands are protected as standing forest.

The idea of improving social and environmental performance in soy production is a seductive one. But the truth is, environmentalists have little leverage against food giants such as Unilever and Archer Daniels Midland. And even if environmental leaders can persuade European and U.S. consumers and companies to buy certified soy only, they freely admit that they have no idea how to influence China, which has repeatedly proven to be impervious to boycott threats and is largely unconcerned with its impact on ecological systems, especially those in other countries.

The more steadfastly environmentalists ignore the macrodrivers of deforestation, the more desperately they propose technical solutions. Over the last few years, the conservation community has become increasingly excited by the prospects of motivating governments to protect their rain forests as a way to receive carbon credits through the Kyoto agreement. The idea is that countries such as Brazil would be able to price and sell their "ecological services" to the world.

Nobody doubts that there is a value to keeping most of the Amazon forest intact. According to one World Bank report, for every hectare of forest destroyed, a hundred dollars in social and environmental benefits are lost — a figure calculated from estimates of particular ecosystem services, from wood products to tourism value to the sequestration of carbon to prevent global warming.[51] And there is no shortage of creative ways to monetize the ecological services provided by standing forests, unobstructed rivers, and clean waterways. The technical know-how exists, and Brazil certainly has a lot of ecological services to supply. The problem is that despite all of the high-minded rhetoric, the United States and Europe have so far been more interested in seeing Brazil service its dictatorship debt than invest in those ecosystem services.

Every few years, environmentalists get excited about yet another technical or tactical approach, from sustainable development projects inside the forest to sustainable soy and cattle production to ecosystem services. But all avoid the drivers of deforestation in Brazil, from its aspirations to economic greatness to the dictatorship debt to land concentration.

Thanks to the efforts of people like rock star Bono, Harvard economist Jeffrey Sachs, and the Jubilee Coalition for debt relief, the last few years would have been an excellent time for environmentalists in Brazil, the United States, and Europe to champion a global solution to Brazil's dictatorship debt. But because environmentalists believe conservation and economic development are separate issues, that thought apparently never crossed their minds.

7.

A cynic might suggest that the lack of enthusiasm for global debt relief among North American conservationists could have something to do with the fact that they have long advanced their own proposal for debt relief — one predicated on the idea that the developed world should swap Brazilian debt for Amazon forest. And therein lies an important difference. Whereas Bono, Sachs, and the debt relief movement point out that the debts owed by developing nations like Brazil were incurred by dictatorships, were paid off several times over, and should thus be considered immoral, the environmental movement treats Brazil's debt as a bargaining chip to be used to secure the Amazon forest without regard for Brazil's economic aspirations. In doing so, environmentalists have done more to define environmentalism as an alien concern than anyone else.

The failure to pursue an approach that protects both the Amazon and Brazil's national interests has stoked concerns that conservation masks a conspiracy to enrich the developed world at Brazil's expense. Today, 75 percent of Brazilians say they fear a foreign country could invade Brazil to take advantage of its natural resources.[52]

The debt-for-nature idea was first proposed in 1984 by the biolo-gist Dr. Thomas Lovejoy, then with the World Wildlife Fund. Lovejoy wrote an opinion piece for the *New York Times* proposing that con-servationists and governments purchase developing world debt and trade it for conservation.[53] At first the Brazilian government opposed debt-for-nature swaps as a threat to their national sovereignty. Al-though the government reversed its official opposition in 1991, debt-for-nature never took off in Brazil, encountering opposition from both the right, which feared encroachments on sovereignty, and the left, which argued that the military-incurred debt should be forgiven outright. To this day, the concept triggers the ire of Brazilians.

The concern about debt-for-nature continues to annoy Lovejoy, now the president of the Heinz Center for Science, Economics and the Environment in Washington, D.C. "It's an irrational response," he told us.

> If you're going to have a sovereignty issue, it should apply equally to debt for equity, where somebody buys discounted debt and uses it to buy an industry in Brazil. Nobody has ever done a debt-for-nature swap when a country didn't want it. Brazil has always had this paranoia and it's odd, but it's endemic.[54]

8.

Sovereignty fears have long driven economic development in the Amazon. Part of what motivated Brazil's military dictatorship's plan to colonize the Amazon was the military's long-standing fear that for-eigners wanted to take control of the region's riches, in particular its biodiversity.[55] The military dictatorship's determination to colonize Brazil's northern and eastern borders was triggered in part by Che Guevara's failed effort to foment revolution in Bolivia in 1968. Fears of internationalization continue to be cultivated by the Brazilian government within the army war college, the Escola Superior de Guerra.[56]

Today, Brazil's military and political elite label anything they don't like as the first step down a slippery slope toward the "internationalization of the Amazon." When members of Congress who had been influenced by conservation biologists suggested that the Amazon be internationalized in 1992, the year of the UN Earth Summit in Rio de Janeiro, Brazil's army generals were quick to announce they would repel any foreign eco-invaders. And periodically Brazilian generals generate headlines promising to resist any foreign efforts to internationalize the Amazon.

Much of the nationalism whipped up by Brazil's military leaders is self-serving, as it motivates support for the government's effort to colonize the Amazon and increases the military's budget. But it is a mistake to dismiss Brazil's concerns, as Lovejoy does, as irrational.

Lovejoy's argument to us — that because Brazil's sovereignty concerns aren't triggered when foreign investors use Brazil's debt to purchase national companies, they shouldn't be triggered when foreign environmentalists use Brazil's debt to conserve the forest — ignores the powerful symbolic, romantic, and geopolitical importance the Amazon has had for Brazilians since the sixteenth century. And given the history of U.S. intervention in Latin America — its support for Brazil's military coup and its help propping up dictators in Argentina, Uruguay, Paraguay, and Chile — not to mention its outright invasion of Iraq in 2003, what's surprising is not that Brazilians have sovereignty concerns but rather that American environmentalists are so dismissive of them.

Lovejoy is an accomplished biologist whose contributions to understanding the Amazon are significant. That Lovejoy has spent years living and working with Brazilians makes it all the more disappointing that he can dismiss their genuine concerns.

Brazilians rightly see their country as a colossus — a regional superpower with the military, economic, and political potential to become a global one. As a country of great artists, architects, diplomats, designers, and engineers, and as a people who speak Portuguese in a region where Spanish dominates, Brazilians justifiably see themselves as special and unique. At the same time, many Brazilians are ashamed of the persistence of widespread poverty, violence, and lawlessness. For many years Brazilians played on a slogan from the Kubitschek era that "Brazil is the country of the future!" by slyly adding, "And always will be!" But it is a joke used to mask their feelings of indignation and resentment.

This stew of national pride and shame results in Brazil's love-hate relationship with the United States. Environmentalists' efforts to reassure Brazilians that their attempts to save the Amazon are in Brazil's best interests not only fail to assuage Brazilian concerns, they trigger Brazil's fear of being patronized, which in part explains why Brazilians get more irritated with do-gooder efforts to save the Amazon than commercial ventures to buy Brazilian corporations.

Brazilians continue to see environmental proposals — such as working with foreigners to conduct a comprehensive inventory of the Amazon's biological assets — as suspicious. Do these foreigners actually care about us? Brazilians ask themselves. Or is this a plot by the CIA or the pharmaceutical companies to rob us of our wealth? Are environmental efforts to win Brazil greenhouse gas credits that can be bought or sold as part of an international treaty on global warming really in Brazil's best interests? Or are they a conspiracy to prevent the country from developing economically? (A question, incidentally, that is being asked by developing countries around the world, from India to China.)

American and European environmentalists have certainly learned to pay lip service to questions of economic development and poverty. "There are many more people living [in the Amazon today than forty years ago], so a reasonable solution is a reasonable quality of life that protects the forest," Lovejoy conceded. "But, if you wait around for [the Brazilians] to fix their economic problems it will be too late."

Little surprise, then, that in their relationships with foreign conservationists, Brazilians, no matter their political orientation, often wonder, "Do you care about us or just our forest?"

9.

The most explicit and influential case for the internationalization of the Amazon was made by renowned conservation biologist John Terborgh, who heads the prestigious Center for Tropical Conservation at Duke University in North Carolina. Terborgh spends much of his 1999 book *Requiem for Nature* arguing that ecological hot spots such as the Amazon should be controlled and patrolled by the United Nations. "In my opinion, the best — perhaps the only — hope lies in the internationalization of nature protection," he writes. "Biodiversity transcends national boundaries and belongs to no one."[57]

Terborgh's opinions are hardly unique. Terborgh is one of the best-known conservation biologists in the world and is influential among American conservation groups. His book was glowingly blurbed by the then-president of the World Wildlife Fund Katherine Fuller; *Collapse* author Jared Diamond; and the famous biologist Edward O. Wilson. *Requiem* had a major influence on American conservation organizations and their funders.[58]

On the one hand, Terborgh stresses that conservation and economics are separate issues, arguing in his book, "Ultimately, nature and biodiversity must be conserved for their own sakes, not because they have present utilitarian value . . . Instead, the fundamental arguments for conserving nature must be spiritual and aesthetic, motivated by feelings that well up from our deepest beings."[59] When we interviewed him he told us, "Poverty alleviation is not what conservation is all about. It's a different enterprise. It's a separate issue."[60] On the other hand, Terborgh repeatedly praises Peru's former dictator Alberto Fujimori for his "enlightened macroeconomic policies."[61]

When we asked Terborgh about the contradiction between his

praise for Fujimori's economic program and his insistence that poverty alleviation and conservation should not be linked, Terborgh became more effusive in his praise for Fujimori. "I think he's the best thing that ever happened to Peru," he told us. "Corruption was very low. What he did was thrilling." Terborgh offered these views to us even though Peru's prosecutors had recently charged Fujimori with stealing $162 million, bribing the media, and murdering his critics.[62, 63]

One gets a sense of how Terborgh would like to see conservation happen in his praise for the way Fujimori — who declared himself dictator in 1992 and dissolved the country's congress and supreme court — ruled Peru. "Active protection of parks requires a top-down approach," he wrote in *Requiem*, "because enforcement is invariably in the hands of police and other armed forces that respond only to orders from their commanders."[64]

Exhibit A in Terborgh's case for internationalizing the Amazon is Manú Park in Peru, where he has worked for three decades. His concern: the handful of indigenous and mixed-race people who he believes shouldn't live in the park. In a chapter ominously titled "The Danger Within," Terborgh laments that the "settlements swarm with children, most of whom survive to adulthood because they have access to public health services provided by the government. Where there are now ten families, in another eighteen years there will be twenty — that is, if nothing else changes, which is unlikely."[65]

Terborgh refers to the situation as a case of overpopulation — a curious description, given that there are only twenty people per square kilometer in Peru, compared to twenty-nine people per square kilometer in the United States, a country that uses per capita far more of the world's resources.[66] What Terborgh is referring to when he speaks of overpopulation is the fact that so many poor and landless peasants are seeking to meet their most basic needs in the forest.

"Is there a solution?" Terborgh asks. "There is." Terborgh proceeds to argue that the government should use cash and other incentives to lure the younger generation of Indians away from the park.

"What I propose is a carefully constructed and voluntary relocation program built on the manifest desire of contacted indigenous groups to acquire goods and an education for their children and to participate in the money economy."[67]

Terborgh laments that his plan might be opposed because "in the present political environment of Perú, indigenous rights are being championed by many groups, including the government . . . In this political climate, proposing the relocation of an indigenous population would be analogous to advocating racial segregation in the United States."[68]

Terborgh spends much of his book attacking "Integrated Conservation and Development Programs" because, in his words, they are "tantamount to social engineering" (in contrast, apparently, to paying Manú Park's Indians to abandon their homes and way of life).[69] "Am I a misanthrope?" Terborgh asks in his book. "I don't think so."[70]

And maybe he isn't. But after reading page after page of Terborgh raging against the encroachments into a piece of nonhuman nature by "squatters," it becomes increasingly clear how one of the world's most influential and prominent conservation biologists would answer the Brazilians' question "Do you care about us or just our forest?"[71]

10.

The way Brazil thinks about the Amazon hasn't changed all that much since 1988, when Brazil's foreign minister declared that "Brazil does not want to transform itself into an ecological reserve for humanity. Our greatest duty is with economic development." Nor has it changed much since 1972, when the interior minister said, "A country that has not yet achieved a minimal standard of living is not in a position to spend its valuable resources protecting the environment."[72]

In early 2007, in response to another intergovernmental report on global warming, Brazil's president Lula da Silva said, "The wealthy countries are very smart, approving protocols, holding big speeches on the need to avoid deforestation, but they already deforested every-thing."[73]

And President Lula da Silva was, of course, right: every one of the world's wealthiest countries long ago destroyed their ancient for-ests in order to develop. Had Brazil developed like the United States — and had the United States developed like Brazil — then today there would exist large organizations of upper-middle-class Brazil-ians funded by Brazilian philanthropists referring to America's forests as the "lungs of the world." Given all of this, the most educated and prosperous inhabitants of the world's wealthiest countries are not in a particularly strong position to pressure Brazil to do more to save the Amazon.

None of this is to deny the ecological reality. The burning of forests, the loss of their role as net absorbers and storage banks of carbon, and the reality of global warming make the increasingly rapid destruction of the Amazon even more alarming than it was back in the mid-1980s, when the Amazon first became appreciated for its biodiversity. Even if we reduced greenhouse gases by 70 per-cent worldwide overnight, the continued destruction of the Amazon would still leave the global climate system in jeopardy.

The political problem remains that the destruction of the Ama-zon is, to the vast majority of Brazilians, far less alarming than the fear of losing their jobs, their life savings, or their lives on the mean streets of Rio de Janeiro. The same is doubly true for the people who eke out a living in the Amazon. Heads of cattle in Mendes's home state of Acre went from 26 to 64 million between 1990 and 2003. Nearly fifteen years later, the Rousseauian dream of extractive reserves as communities of peasants, rubber tappers, and Indians working and living in harmony with nature has faded. Every extrac-tive reserve in Acre is today grazing cattle. When *Veja* magazine re-porter Leonardo Coutinho visited the Chico Mendes Extractive Re-serve in Acre, he discovered that even Chico Mendes's widow had a herd of them.[74]

Instead of grappling with the macroeconomic forces driving the forest's destruction — and articulating a national vision that speaks to the aspirations of the Brazilian people — environmentalists inside and outside of Brazil have spent fifteen years designing and promoting technical pilot projects at the grassroots and reinforcing the sense that protecting the Amazon should be done for environmental reasons. They demand that the government do more to enforce the law — and refuse to deal with the reasons behind the government's lack of police work in the Amazon. As a consequence, environmentalists both in the government and in nongovernmental organizations have seen their influence since Mendes's death decline, not rise.

For too long, environmentalists have believed their cause — protecting nature — to be so transparently right that they have thought little about their failure to appeal to deeply held national aspirations. It has thus been inconceivable to environmentalists that they might do more to protect the Amazon by addressing, for example, Brazil's desire to be an agricultural superpower, a UN Security Council member, or an industrial leader in biotechnology.

Until the world's wealthiest countries seriously support Brazil's goals for itself, the colossus of the south will have neither the means nor the motives to save the Amazon. In a country where one out of five people goes to bed hungry every night, where drug traffickers control 100,000-person *favelas* in Rio, and where middle-class professionals fear for their financial security and their personal safety, saving the Amazon simply will never become a top concern. Protecting the Amazon and ensuring future rainfall is in Brazil's long-term economic interests, to be sure, but as long as its short-term and long-term interests collide, the forest will continue to burn.

If our answer to the Brazilians' question "Do you care about us or just our forest?" is indeed "We care about you," then we must do something for Brazil before we can do anything for the Amazon.

3

INTERESTS
WITHIN INTERESTS

I N 1987, CIVIL RIGHTS activists working at the United Church of Christ conducted a study that concluded that toxic waste facilities were more likely to be located near minority communities than white communities.[1] The study generated national headlines, searing into the minds of millions of Americans the notion that corporations were deliberately targeting minority communities with their hazardous waste.

Four years later, at the 1991 People of Color Environmental Leadership Summit, civil rights leaders described the disproportionate impact of pollution on poor American communities not only as racist but also as genocidal. The environmental justice movement's founding document, issued at the conference, stated, "Environmental justice considers governmental acts of environmental injustice a violation of international law, the Universal Declaration on Human Rights, and the United Nations Convention on Genocide."[2] The UN Convention explicitly defines genocide as intentional acts "to destroy, in whole or in part, a national, ethnical, racial or religious group."[3]

Even though there is no evidence that the U.S. government or corporations have intentionally been trying to destroy communities of color through pollution, environmental justice (EJ) activists have rarely showed much hesitation in playing the genocide card. "This is genocide at its finest," declared Pat Bryant, a prominent grassroots EJ leader, in testimony before Congress in 1993.[4]

In the years that followed the United Church of Christ report, a raft of other studies by environmental justice activists, sympathetic journalists, and left-leaning academics made similar claims, and the terms *environmental racism* and *environmental justice* soon entered the political lexicon. The new movement grafted the race-based politics of the civil rights movement onto the place-based politics of grassroots environmentalism in a hybrid that many hoped would create a more expansive, inclusive, and powerful basis upon which to advocate for both social justice and environmental protection.

Along the way, the movement served most of its original protagonists well. Today, when national environmental leaders are criticized for being out of touch, elitist, or irrelevant, they point to environmental justice activists in order to demonstrate the inclusiveness and relevance of environmentalism to poor, nonwhite Americans.

The new movement also brought its founders to great prominence. The liberal establishment quickly embraced Benjamin Chavis Jr., the deputy director of the United Church of Christ's commission that authored the first report. In 1992 he was appointed to President Clinton's White House transition team. Soon thereafter he was named head of the NAACP. And a report commissioned by EJ activists and their funders revealed that, over just four years, between 1996 and 1999, funding for environmental justice activities went from $27 million a year to $49 million a year.[5]

Less clear, however, is what the environmental justice movement has done for those whom it claims to represent: the members of low-income communities of color. The movement has won no significant new environmental laws or any major civil rights legal challenges. To the extent that pollution affecting low-income communities has declined, it has been thanks to the enforcement of environmental laws passed in the 1950s, '60s, and '70s — long before environmental justice advocates declared pollution a kind of genocide. Meanwhile, air and water pollution remain, unsurprisingly, far lower priorities for poor Americans of every color than jobs, health care, and public safety — problems that the EJ movement, despite its claims to expansiveness, largely ignores.

Poor Americans of all races, and poor Americans of color in particular, disproportionately suffer from social ills of every kind. But toxic waste and air pollution are far from being the most serious threats to their health and well-being. Moreover, the old narratives of intentional discrimination fail to explain or address these disparities. Disproportionate environmental health outcomes can no more be reduced to intentional discrimination than can disproportionate economic and educational outcomes. They are due to a larger and more complex set of historic, economic, and social causes.

In this way, environmental justice symbolizes something larger than the failure of environmentalism to address the most important issues facing poor communities and communities of color. Race-based environmentalism is a case study in how the *anti*-ecological logic of interest-group liberalism constricts rather than expands our concerns, and becomes complaint based — all of which make it poorly suited to dealing with large, complex, and deeply rooted social and ecological problems.

1.

On December 19, 1993, Christopher Foreman Jr., a senior fellow at the Brookings Institution in Washington, D.C., read a *New York Times* story about President Clinton's intention to issue an executive order directing federal agencies to set up internal processes to guarantee environmental justice. Over the next few days Foreman looked into the issue and was shocked by what he read. "A lot of what I found in the EJ literature was essentially, 'People of color are being poisoned, *en masse*, willy-nilly, by corporate interests,' and so on. I thought, 'That's outrageous! How can we not know about that?'"[6]

Foreman had recently completed his second book, the well-received *Plagues, Products and Politics: Emergent Public Health Haz-*

ards and National Policymaking, which earned him the reputation of a scholar capable of bringing a critical eye to various fields, from epidemiology to sociology to politics. Foreman was particularly interested in the intersection of health, race, and class, and he decided to write a book on the subject of environmental justice. "I was taking for granted all of the claims of the EJ community," Foreman admitted when he spoke with us. "I quickly discovered that a lot of the things I was reading in the environmental literature were not squaring very well with claims from the EJ community."[7]

Foreman spent the next five years carefully sifting through the evidence. Today, *The Promise and Peril of Environmental Justice,* published in 1998, remains one of the most important books ever written on environmental justice. At just 191 pages long, *Promise and Peril* sheds far more light on the politics of race, class, and environment than books on environmental justice three times its length. In the book, Foreman concludes, "Empirical support for claims of disproportionate pollution impacts and discriminatory regulatory enforcement is actually much weaker than environmental justice advocates usually admit." As a result, the EJ movement "probably directs community attention away from those problems posing the greatest risks and *may therefore have the ironic effect of undermining public health in precisely those communities it endeavors to help.*"[8]

Foreman's central insights and predictions held up well in the years after he published *Promise and Peril.* He concluded that EJ advocates would continue to fail to win lawsuits brought under civil rights laws, which were not designed to deal with pollution problems. And Foreman predicted that there would be no further action by Congress or the EPA on environmental justice questions. "In the foreseeable future, the EPA and Congress will probably lack strong scientific justification for major policy change on this front."[9]

Most incisive critics of environmental justice research and politics are, like Foreman, scientists concerned with public health. They are also liberals who are deeply sympathetic to the plight of poor communities. They are not, as EJ activists often claim, industry scientists greenwashing toxic polluters. Foreman, who describes himself as a progressive Democrat, repeatedly makes clear his concern about the power of corporations to manipulate the political process. "The most fundamental systemic challenge" for activists seeking to empower local communities, Foreman writes, "is the generally superior position (in both participatory resources and incentives) of vested economic interests."[10] For Foreman, the superior position of corporations makes it all the more important that communities focus on the greatest threats to their health.

2.

In 1992, two reporters with the *National Law Journal* conducted and published the results of a study they had conducted of the government's treatment of companies handling toxic waste in white communities versus minority communities. Marianne Lavelle and Marcia Coyle concluded that "there is a racial divide in the way the U.S. government cleans up toxic waste sites and punishes polluters. White communities see faster action, better results and stiffer penalties than in communities where blacks, Hispanics and other minorities live."[11]

Their finding sent shock waves through the environmental establishment and seemed to provide ironclad proof of the suspicion that had been on the minds of civil rights and EJ activists since the 1987 report on race and toxic waste. The study came out in September 1992, six weeks before Bill Clinton was elected president. The timing helped EJ activists make the case for becoming part of the Clinton transition team.

But the *NLJ* study, which was neither conducted by scholars familiar with statistical analysis nor peer-reviewed, was immediately challenged by Mary Bryant in an article called "Unequal Justice? Lies, Damn Lies, and Statistics Revisited."[12] A few years later, policy analyst and attorney Mark Atlas conducted a statistical analysis of the data for a 1997 article titled "The Contaminated Grassy Knoll: Searching for Environmental Justice Conspiracies in Environmental Enforcement."[13] Atlas found no evidence that environmental law violations in communities of color or low-income communities tended to be penalized less than violations elsewhere.

> With respect to race characteristics, my analysis consistently revealed a modest relationship with penalties, but it was in the opposite direction than that claimed by *NLJ*. Penalties tended to be higher as the presence of minorities increased. The factors that influence penalties the most, though, were the characteristics of the cases.[14]

In the end, Atlas concluded that what determined how much facilities were penalized had to do with the facts of the case, not prejudice either for or against facilities in communities of color.[15]

Atlas's definitive deconstruction of the *NLJ* study came out in 1997, but EJ activists and academics to this day cite the study as proof that the government is prejudiced against communities of color. EJ scholar Robert Bullard, a sociologist at Clark University and the movement's leading intellectual, cited the study in his 2005 Sierra Club book *The Quest for Environmental Justice*, but he does not mention, much less refute, the Bryant or Atlas critiques.[16]

Almost anyone who endeavors to make the long slog through two decades of environmental justice writing can understand why Foreman concluded that EJ research has always been "merely instruments of the movement rather than its cause." What has emerged in the public health literature over the years debunks the theory that corporate villains are deliberately poisoning communities of color — thereby committing genocide — by siting polluting facilities in poor and minority neighborhoods.

In 2003, Pamela Davidson, a professor at George Washington University and a specialist in data collection and analysis, conducted a comprehensive review of the scientific literature on race, class, pollution, and public health and concluded the following:

> Despite the early consensus, close to two decades of environmental justice research has failed to produce conclusive evidence in support of the siting hypothesis, resulting in a divisive field of study . . . In focusing on siting biases and distributional disparities, environmental justice researchers have failed to produce clear and consistent evidence. As such, there is no conclusive evidence of a widespread targeting of racial and ethnic neighborhoods for the siting of noxious environmental sites.[17]

To be sure, poor communities have long been subjected to more pollution than wealthier ones, and communities of color have been subjected to more pollution than white ones. But these communities have not been targeted for pollution because of the race of their inhabitants. Corporate executives locate their facilities in neighborhoods for a variety of reasons having to do with zoning regulations, easy access to transportation, low land prices, the likelihood of residents to effectively resist in a not-in-my-backyard (NIMBY) campaign, and historic development, which was, indeed, often discriminatory. But executives and government officials are not siting facilities in those neighborhoods out of racism. Labeling the disproportionate impact of pollution on Americans of color as either racist or genocidal serves only to distort the community's understanding of the problem and alienate potential allies.

3.

Not only do environmental justice activists see conspiracies that don't exist, they also ignore the ones that do. Powerful forces have conspired to poison Americans of color — and whites, too. But they are not the waste, chemical, or oil firms. They are the food, beverage, alcohol, and tobacco companies.[18]

EJ activists attack those who point out that pollution is not a top cause of diseases such as cancer,[19] so it's important to state one thing right off the bat: diet, tobacco, and alcohol don't have a *slightly* greater impact on health; they have an *exponentially* greater impact. Diet and tobacco alone are estimated to have a *thirty times higher* impact than pollution on people's health. A 1996 Harvard Center for Cancer Prevention report concluded that tobacco caused 30 percent of all cancer deaths. Poor diet and obesity caused an additional 30 percent. The report estimated that occupational factors caused 5 percent of cancer deaths, while environmental pollution caused just 2 percent.[20]

Public health researchers use the word *environment* in a tellingly different way than environmentalists use it. Most epidemiologists and cancer researchers say that the environment is responsible for a large percentage of cancer risks. However, the environment they are referring to *includes* behavioral and lifestyle factors such as diet, tobacco, alcohol, and childbearing, in addition to infectious agents, radiation, and traditional pollutants in the air, water, soil, and consumer products.

The EJ community's indifference to smoking and obesity is particularly egregious given that they are two of the leading preventable causes of death in the United States. Not only that, smoking, which causes 87 percent of lung cancers, has a racially disproportionate impact. Lung cancer kills more African Americans than any other cancer. Black men are roughly 50 percent more likely to develop lung cancer than white men, and more than a third more likely to die from it.[21] And it is the poor who are most likely to smoke. While 22 percent of Americans at or above the poverty line smoked, 34 percent below it did.[22]

The irony of the false accusation that polluting corporations deliberately target communities of color is that alcohol, food, and tobacco companies truly do target communities of color. They invest billions in market research and advertising aimed at branding menthol cigarettes as cool in the eyes of black youth and positioning hard liquor and beer as sex symbols for Latinos.

Of course, EJ leaders not only don't work on smoking, diet, or other lifestyle issues, they tend to dismiss anyone who points out that they have far greater health impacts than environmental issues. When we asked Luke Cole, an EJ attorney, why EJ activists work on pollution but not tobacco or diet, he said:

> I think it's a bogus argument. The environmental hazards are the vehicle to bring people together. Once they come together, they can look at other things in their communities. Anytime you have an organizing tool to get people previously unorganized to come together and realize their collective power, it's a good thing.[23]

All of which sounds good in principle, but there is simply no evidence that EJ activism leads to reduced smoking, better eating, less drinking, or other kinds of health-protective behaviors. It may be that EJ activism has the opposite effect, communicating to community residents that they should be more concerned about pollution than the far more serious threats of smoking, drinking, and a poor diet. And even if we accept the presumption that all community organizing is good, then why not organize around the most significant and salient public health issues?[24]

EJ leaders know that pollution is a far smaller threat than smoking, alcohol, and diet, which is why they have long resisted using the established public health tool of risk assessment to guide public policy. "The question of environmental justice is not anchored in a debate about whether decision makers should tinker with risk management," wrote Bullard.[25] One rather earnest writer, baffled by Bullard's resistance to risk analysis, pointed out:

> *A strategy that emphasized attacking the largest and most easily reduced risks first would appear to represent a major gain for minority communities.* To the extent that such communities bear unusually high risks as a result of past discrimination or other factors, a risk-based approach would . . . give highest priority

to attacking precisely the kinds of problems that most concern Bullard.[26]

But Bullard's main concern is not health risks facing minority communities; it is the overlap of pollution and racial politics. Little wonder then that Bullard is so alarmed by an approach that would represent *all* health risks facing people of color — not just those he defines as environmental.

4.

EJ leaders frequently insist that environmental issues are among the top concerns of people of color and poor Americans.[27] They believe that air pollution, waste facilities, lead paint, and other environmental hazards are an effective way to organize politically. But in focus groups we have held with people of color from around the country, including people from disproportionately polluted communities such as Richmond and Oakland, California, not once have we heard black or Latino Americans name pollution, toxic waste, lead paint, or asthma as one of the main problems facing their community.

In focus groups, when we ask, "How are things going around here?" inevitably what we hear people talk about are unemployment, crime, schools, and health care. Given what a relatively low-risk health factor pollution is, Americans of color are rightly more concerned about nonenvironmental than environmental issues facing their communities.

A cursory read of the polling data shows that communities of color care about pollution about as much as white communities — which is to say, *not very strongly.* When asked to name the main issues facing their communities, people give a whole set of concerns from jobs and health care to education and crime to terrorism and war before they name pollution, toxic waste, lead paint, or the environment.

When we asked Paul Mohai of the University of Michigan, one of the most prominent researchers of public opinion on EJ issues, about his claims that EJ concerns are highly salient, he told us that on his survey, "seventy percent of both blacks and whites mentioned some sort of pollution issue, and 30 to 35 percent mentioned nature preservation. The other categories [including global warming, acid rain, ozone depletion] were smaller. One of the biggest surprises for me was that there wasn't much difference between blacks and whites."[28]

When we asked him what his work had accomplished, Mohai said, "I think that where I see one contribution is dispelling what I consider the mythology that African Americans are unconcerned about environmental issues. If you really look at the data, they're just as active as white Americans." But when we pressed Mohai as to whether environmental issues were a *high* priority for black Americans — not just whether they were an *equal* priority when compared with white Americans — he said he did not know.

Like most EJ research, Mohai's has been aimed at making a political point — namely, that people of color care about the environment just as much or more than whites. "Originally, we were interested in making a comparison between the relative importance of the environment with people," Mohai explained. "To be honest, we didn't get very far in analyzing the answers. Our own research was becoming more focused on EJ."[29]

Like environmentalists who boast that 70 to 80 percent of the country believes global warming is important and that action should be taken, Mohai and his colleague Bunyan Bryant, also at the University of Michigan, have been driven by the desire to show that communities of color want action taken on pollution. What they fail to mention is that when Americans are told about virtually *any* problem they are familiar or unfamiliar with, slightly concerned or very concerned about, from pornography to violence on television to liquor stores to port security, they say they want to see action taken by the government to fix it.

5.

In recent years, some environmental justice groups have increasingly focused on disproportionate rates of asthma experienced by Americans of color. Understanding asthma is a fascinating way to understand how complex racial health disparities actually are.

The basics are these: emergency room visits and hospitalization rates for asthma are higher for black children than white children — and black children are four to six times more likely to die from asthma than white children.[30] Black Americans, as the data below show, are somewhat more likely to suffer from asthma than white Americans. Hispanics, by contrast, are *less* likely to suffer from asthma than whites.

But where the real disparities show up is in the second column, which tracks emergency room visits for asthma. Both black Americans and Hispanics have much higher rates of emergency room visits than do white Americans — even though black Americans have higher overall rates of asthma, and Hispanics have lower overall rates of asthma, than do white Americans.

Asthma is a highly treatable disease. Outpatient care and drug therapies allow most asthma sufferers who have access to those services to avoid emergency rooms. A substantial proportion of those who do end up in emergency rooms for asthma are there because they lack the access to nonemergency care that would allow them to stay out of emergency rooms.

Asthma by Race, 2002, Centers for Disease Control[31]

Race	% reporting asthma	% reporting ER visit for asthma in the last year
All Americans	7.2	18.4
American Indians	11.6	20.4
Blacks	9.3	37.2
Whites	7.6	14.5
Hispanics	5	26
Asians	2.9	18.8

Disparities in emergency room visits form the primary basis for the claim that racial disparities in exposure to air pollution are the cause of the urban asthma epidemic. But a closer examination of the data suggests that lack of access to appropriate treatment and health care is at least as significant a part of the problem facing African American and Hispanic asthma sufferers as is pollution. Clearly, anybody who cares about asthma in those communities must address the issue of access to health care.

6.

In 2001, Geoffrey Canada of the Harlem Children's Zone joined forces with the Harlem Hospital Center and Columbia University's Mailman School of Public Health to identify the various causes behind his community's asthma epidemic. Canada and his colleagues decided they would test — and then treat — every single child in central Harlem for asthma.[32]

What Canada did stands in stark contrast to EJ's narrow race- and environment-based model of thinking and organizing. Rather than deciding in advance to focus only on those causes of asthma that environmentalists consider environmental (e.g., air pollution), Canada did something that in retrospect seems obvious: he attacked every cause of the asthma epidemic he could find. In the process, Canada ended up creating a kind of social safety net for Harlem's children that stands as a model not just for overcoming asthma but also for overcoming poverty, joblessness, and hopelessness.

Canada, who authored an acclaimed memoir, *Fist Stick Knife Gun*, about growing up as a street tough, is famous for setting his sights high — to some, impossibly high. "When I first met this group, I thought that they were nuts to think that they could test every child in an entire community, and then provide services to all of the ones who have asthma," said Mary Northridge, an associate professor at Columbia's Mailman School and editor in chief of the *American Journal of Public Health*. "It's an enormous undertaking."[33]

What Canada and his colleagues found was astonishing: more than 25 percent of the children they tested had asthma.[34] It was one of the highest rates ever documented for a single neighborhood. Their research was consistent with the results of other studies, which found asthma to be caused by *many* factors not just air pollution, from breathing secondhand smoke to breathing indoor air pollution to emotional and physical stress to the lack of health care. The Harlem study found that children with asthma are 50 percent more likely to live with someone who smokes than nonasthmatic children. They discovered that the causes were themselves *over*determined: there are causes within causes. Indoor air pollution is caused by everything from cockroach feces to mold to dust mites. And even those parents who get medical care for their children often do not understand their children's condition well enough to help them take their medication properly. What's needed is not just access to health care, but to attentive health care. Canada and his team sought to tunnel through all of these obstacles and focused on all of the causes at once.[35]

Canada and the Harlem Children's Zone are justly famous for their intensive and complex interventions around economic development, violence prevention, and health. In the asthma project, Canada and his team arranged for doctors, social workers, nurses, and others to repeatedly visit families and make typical health interventions — such as accompanying them to medical appointments and helping administer medicines to their children — as well as atypical interventions, such as repairing broken windows and holes in walls and replacing dusty carpets and furniture.[36]

Canada's team even engaged volunteer lawyers to threaten landlords into making housing improvements.[37] One family had complained for months to a landlord about abysmal conditions, even pointing out that a rodent mother had given birth in their bedroom's dresser drawers, to little avail. The mold "looked like hair coming off the wall," the social worker involved in the project told the *New York Times*. "Then, after lawyers got involved, action was pursued immediately."

Canada's program was wildly successful. At the beginning, 35 percent of the children reported going to the emergency room or making emergency trips to the doctor's office. Just eighteen months later, that number had been reduced to 8 percent. Overnight hospital stays were reduced from 8 percent to zero. And the number of children who reported missing school because of asthma fell from 23 to 8 percent. The program — and its multifactorial approach to asthma, which deals not just with kids but also with their parents, teachers, relatives, and others — is today considered by the CDC to be a model for other childhood illnesses, from obesity to diabetes. The upfront investment, which was paid for by the Robin Hood Foundation, was $1,700 per child. With a one-night hospital stay in Manhattan costing more than $4,500, the program was incredibly cost-effective.[38]

Canada's asthma program is just one aspect of the Harlem Children's Zone, a project that serves 8,600 low-income children in a sixty-block area. It also has antiviolence, tutoring, mentoring, parenting, preschool, and educational initiatives. It works because it covers children from birth to college with a range of services that simultaneously address multiple causes of multiple problems. The results are impressive. Parent graduates of Baby College sing and read more often to their children than nongraduates, and their children are more likely to get immunizations. And in one school year, the Zone's Promise Academy helped kindergarteners go from 11 to 80 percent "school ready."[39]

In thinking ecologically and behaving pragmatically, the Harlem Children's Zone invented a kind of social safety net that all Americans should have.

7.

Given the strong relationship between asthma and poor or absent health care, it is understandable that the Harlem Children's Zone focused so much on improving treatment — and all the more inexplicable that the local environmental justice group WE ACT (West Harlem Environmental Action, Inc.), which is routinely held up as a model EJ group, has steadfastly ignored the role played by housing and inadequate health care.

WE ACT is often cited as evidence that the EJ community is expanding its focus beyond toxins. But in contrast to the Harlem Children's Zone's multifactorial approach, WE ACT's advocacy focuses on a single source of pollution: New York Metropolitan Transit Authority's diesel buses. WE ACT sued the MTA under Title VI of the 1964 Civil Rights Act, claiming that its siting of bus depots in Harlem was racist. The suit, like every other EJ suit filed under Title VI, failed.

Attempting to piggyback onto the publicity of a comprehensive 2001 study of asthma in Harlem, WE ACT's environmental health director Swati Prakash sent out a press release saying, "We need to concentrate on the preventable exposures that may contribute to childhood asthma, and air pollution is at the top of that list."[40] But contrary to what Prakash claimed, there is simply no evidence that air pollution in general, or diesel exhaust from buses in particular, is the number-one preventable cause of childhood asthma in the area. The study offered no such ranking at all. Nor did WE ACT's executive director have any evidence to claim that "health disparities between communities in New York City continue to increase due to the concentration of polluting facilities in certain areas." WE ACT's press release was based not on the new asthma research but on the same talking points it had created to attack the MTA three years earlier.

We asked WE ACT's cofounder Vernice Miller-Travis to describe the main concerns of Harlem residents. She told us it was its disproportionate asthma.[41] But when asked about their top concerns, Harlem residents name jobs, crime, health care, housing — the well-known list of ills that beset urban communities — not asthma. It is these issues — not just asthma caused by diesel buses — that concern the Harlem Children's Zone. It is for this reason that the HCZ deals with housing, health care, daycare, parenting classes, and violence prevention. Many of these services, such as health care and housing advocacy, reduce childhood asthma and are part of the HCZ's multifactorial approach. But others are just part of dealing with the issues that matter most to the people of Harlem.

In contrast to the Harlem Children's Zone and against the rhetoric of EJ activists and academics who claim their movement "incorporates the aims of other social movements,"[42] WE ACT, like EJ groups generally, is focused on a single problem (asthma) and a single cause (diesel buses). It is a strategy that goes against the science of public health, which today focuses on the risks from synergies among multiple risk factors not single threats,[43] and contradicts the claims of EJ's academic boosters who believe that the movement has expanded the definition of the environment and improved the lives of poor people of color.

Following the findings of the best available social science on human motivation, Canada is focused on understanding and building community and individual assets, not deficits; strengths, not weaknesses; aspirations, not resentments. He understands the importance of being conceptually expansive in his approach. In contrast to WE ACT, the Harlem Children's Zone doesn't fetishize a single cause of asthma, nor does it bank everything on a single program.

The problem is not, as some critics of environmental justice have claimed, that environmental justice advocates and scholars have an agenda. After all, *everyone* has an agenda, whether he or she is honest about it or not. To study something is to value it. Public health values the study of health. Public health, which includes both the so-called natural sciences of biology and genetics as well as the so-called social sciences of sociology and psychology, values improving health outcomes. When there are disagreements within the field of public health, they are over how best to improve health outcomes.

The problem is that EJ researchers want all the credibility of doing public health research and none of the obligations. Epidemiology begins by understanding a disease and then proceeds to search for its causes. This approach has been used successfully to guide the development of public health strategies on everything from AIDS to avian flu to cancer. The environmental justice approach begins with potential environmental threats, such as incinerators, waste facilities, refineries (but not smoking, diet, stress, poor health care, or housing) and aims to prove that they have negative health effects.

That is because environmental justice is primarily interested in advancing and merging two discourses: one about racial prejudice and inequality and the other about environmental pollution. The goal is not to improve health outcomes for the poor or nonwhite communities. The result is that environmental justice advocates tend to be interested in the health and well-being of communities of color only insofar as their research demonstrates that health problems are both caused by pollution and greater among minorities. Ironically, environmental justice activists often end up focusing on concerns that are neither particularly significant nor particularly salient to the communities in question.

8.

The greatest strides in reducing exposure to pollution in communities of color — and everywhere else in America, for that matter — have been made not through the race-based strategies advocated by environmental justice activists but rather through the enforcement of national environmental laws established long before the environmental justice movement was a glimmer in Ben Chavis's eye. Before anyone was even talking about disproportionate pollution impacts, laws such as the Clean Air Act were passed with the intention of protecting *all* Americans. And that law has, for the most part, worked: air quality has vastly improved for all communities. The total annual emission of 188 regulated toxins declined by 36 percent in thirty years.[44] Similarly, in the case of lead poisoning, where there is a disproportionate impact in terms of race and class, most of the most important government actions to reduce lead poisoning, such as banning lead paint in 1972, phasing out leaded gasoline in 1986, and requiring real estate agents and landlords to disclose residential lead paint, passed without the existence of an environmental justice movement.[45]

It is more than a bit ironic then that the environmental justice movement came into being accusing national environmental organizations of racism. On January 16, 1990, a group of prominent civil rights leaders sent an open letter to the heads of the country's ten largest environmental organizations accusing them of racist hiring practices and demanding that they hire more people of color. Two weeks later, the *New York Times* ran a story titled "Environmental Groups Told They Are Racists in Hiring."[46] The letter generated a huge amount of publicity and attention for the civil rights leaders and their accusations of environmental racism.

Seeing how much attention the letter generated, an activist in Albuquerque, New Mexico, Richard Moore, organized and publicized his own open letter two months later. "Industrial and municipal dumps are intentionally placed in communities of color," he charged, and environmental organizations "play an equal role in the disruption of our communities" as mining companies, the military, and toxic dump operators do.

And Moore reiterated the accusation that environmental leaders are racists.

> Racism is a root cause of your inaction around addressing environmental problems in our communities . . . We again call upon you to cease operations in communities of color within 60 days, until you have hired leaders from these communities to the extent that they make up between 35–40 percent of your entire staff.[47]

Then, a few weeks later, Moore and a smaller group of signers sent a letter to a different group of environmental organizations, most of them smaller than the national groups, asking that within the year they ensure that *50 percent* of the boards of directors and staffs be composed of nonwhites.[48]

Given how white environmental staffs were, meeting Moore's demand that 50 percent of the staff of environmental organizations be nonwhite within one year would have required the groups to *fire close to half of their employees for being white* — a move that would likely have raised eyebrows at the U.S. Department of Justice's Office of Civil Rights, which is charged with enforcing nondiscrimination in employment. It was a rather ironic demand coming from groups ostensibly concerned about not just racism but also workers' rights.

Were environmental organizations really discriminating against people of color? Neither Moore nor other civil rights leaders ever offered any evidence. They had no whistleblowers, no memos, no history of lawsuits or settlements to back up their accusations. But then, no evidence was required. The whiteness of environmental organizations was all the proof anyone needed. That's because few things are more terrifying to a white liberal than being accused of racism — and few things are more threatening to fundraising than negative publicity. The civil rights activists knew that environmental leaders and philanthropists would want them to put away the race card as quickly and as quietly as possible.

They also knew that within environmental organizations, white guilt was far more prevalent than white racism. Indeed, the 1990 letters are best understood as an extremely effective public relations campaign aimed at defining white environmental leaders and their organizations as racists. It was a strategy grounded in the understanding that the charge alone would panic environmental executives and their funders — and result in the increased power of those making the charge. "The letters were really about who had the power within the environmental movement," Moore explained, "and how it was going to be used."[49]

The media campaign created momentum for the 1991 People of

Color Environmental Leadership Summit, where environmental leaders defensively insisted, "We are not the enemy," and duly promised to diversify their staffs.[50] It was shortly after the summit that mainstream environmental, civil rights, and environmental justice leaders achieved a tacit accommodation: environmental groups would hire more people of color to work on environmental justice, environmental foundations would send money to environmental justice groups, and civil rights and environmental justice groups would lay off the high-profile accusations of racism.

9.

Accommodating the demands of the nascent environmental justice movement in the late eighties and early nineties was an easy call for the white liberal leadership of the national environmental movement and the philanthropies that supported them. Nobody likes to be accused of racism, and environmental funders and leaders had little to lose once it became clear that the criticism leveled by environmental justice activists would not go beyond having to hire more nonwhite staff and focusing on pollution in poor, nonwhite communities.

But when one steps away from the rhetoric of environmental racism and genocide, there is an oddly dissonant feeling to the whole thing.

Like environmentalists in Brazil who advocate sustainable development in the forest while turning a blind eye to the *favelas* of São Paulo, environmental justice advocates, with the blessings of mainstream environmentalists, focus on diesel buses, oil refineries, and asthma in poor communities that are faced with much larger crises, from joblessness to poor schools to the lack of affordable health care.

While there is little evidence of a conspiracy to poison nonwhite Americans with toxic pollutants, Americans living in low-income, predominantly nonwhite communities disproportionately suffer a variety of poor health outcomes. In a winner-take-all economy, such outcomes are almost inevitable. Poor Americans are more likely to live in polluted urban areas and in close proximity to polluting industrial facilities, highways, and bus and truck routes. They are also much more likely to lack adequate health care and housing, to work in industries with high occupational health hazards, and to suffer disproportionately from a variety of nonenvironmental health ills. They have higher infant mortality rates; lower life expectancy; higher levels of substance abuse, domestic violence, and incarceration; and higher rates of obesity, diabetes, and heart disease.

To be poor in America is to be unhealthy, and there are a multiplicity of causes. And therein lies the problem. Poor communities, particularly poor, predominantly nonwhite communities, are confronted by a staggering array of social ills. Jobs are scarce, basic community services such as supermarkets are nonexistent, health care is poor, schools are falling down, gangs and drug-related crime are ubiquitous. The environmental justice movement ignores all of these concerns because, like the mainstream environmental movement, it sees them as separate from the environment.

Despite their rhetoric of expansiveness, environmental justice advocates have made environmentalism smaller, not larger. It is a category within a category, narrowly focusing on specific environmental issues and addressing neither the primary public health issues confronted by communities of color nor the central issues of economic opportunity, education, and social mobility that lie at the heart of poverty in America.

The environmental justice movement has failed to provide a meaningful and compelling agenda because it continues to see the environment as a thing separate and distinct from everything else. Why else would environmental justice advocates direct their efforts toward reducing exposure to toxic chemicals from refineries but *not* from cigarettes? Why else would they focus on eliminating diesel bus emissions that contribute to childhood asthma but *not* improving dilapidated housing that contributes at least as much to the same epidemic?

If environmentalists see the world through green-tinted glasses that filter out the social context for the outcomes they desire, environmental justice advocates wear blinders. In focusing narrowly on finding racial disparities in exposure to environmental pollution, they ignore virtually everything that really matters to the communities on whose behalf they claim to speak.

Whether in São Paulo or Harlem, environmentalists will remain small and irrelevant so long as they ignore the central importance of prosperity to both community health and ecological concern. The legendary labor organizer Saul Alinsky taught that effective organizers "start where people are at." And where most Brazilians and most residents of Harlem are "at" are communities faced with needs that simply take precedence over environmental pollution or the relatively insignificant health threats that environmental pollution causes.

Today there are signs that some local environmental justice organizations around the country are evolving into something more expansive, something that doesn't insist on placing pollution at the center of their politics. Groups that were once focused strictly on air pollution are today advocating jobs, health care, and kinds of urban development that lead to livable communities. These post–environmental justice organizations, most headed by younger leadership, are in various stages of creating something more expensive.

Many environmentalists and environmental justice advocates argue that economic growth and development are not inconsistent with sustainability and environmental health, and in this they are correct. There is no reason that Brazilians and Harlem residents can't get their various material and postmaterial needs met in ways that also result in a better environment. There are no shortage of strategies that could lead to a better, more prosperous life for residents of Harlem or São Paulo *and* have the significant ancillary benefits of cleaning the air in Harlem and protecting the Amazon in Brazil. But for those strategies to amount to a more powerful and relevant politics, the dream of a better, more prosperous life must be at the center, not at the margins.

4

THE PREJUDICE OF PLACE

A FEW DAYS BEFORE Christmas 2005, Robert Kennedy Jr., one of America's most famous environmentalists, published an op-ed piece in the *New York Times* attacking Cape Wind, a large wind-energy project slated to be built six miles off the coast of Cape Cod, in Nantucket Sound, Massachusetts. Famous for his crusades against the Bush administration's "crimes against nature," Kennedy, a senior attorney with the Natural Resources Defense Council, had in 2004 and 2005 become increasingly outspoken about global warming, so his attack on one of the largest clean-energy developments in the world struck many as curious.

In his op-ed piece, Kennedy made a variety of arguments familiar to anyone who has observed a not-in-my-backyard (NIMBY) campaign in action. He began with a statement of his best intentions. "As an environmentalist," Kennedy wrote, "I support wind power, including wind power on the high seas."[1] Kennedy explained that, in fact, he was a strong proponent of wind power in other places — just not in Nantucket Sound.

And so began the now-familiar litany of NIMBY complaints about inappropriate development carefully wrapped in grand ecological narrative. Nantucket Sound, Kennedy explained, is unique, unspoiled, and fragile. The roar of the wind turbines would overwhelm the senses of those who were able to truly appreciate the Sound. The navigation lights on the towers would "steal the stars." Mineral oil housed at the site could contaminate the Sound. The blades on the wind turbines would kill thousands of songbirds. A web of underground cables would make it impossible for local commercial fishermen to earn a living.

Kennedy implied that he was motivated by nothing more than his desire to conserve a treasured natural place no less precious than Yosemite National Park, protect the commons from rapacious energy companies, and preserve the need for "all of us to periodically experience wilderness to renew our spirits and reconnect ourselves to the common history of our nation, humanity and to God."

But among the facts Kennedy omitted from his piece was a rather personal and inconvenient one: the fact that the proposed development would be visible from the Kennedy family compound in Hyannis Port.

1.

Kennedy's employer, the Natural Resources Defense Council, was founded on the premise that strict adherence to the scientific facts is central to the success of environmentalism, and Kennedy himself spent much of 2005 blaming the media for failing to properly inform the public of the basic facts of global warming. But in the case of his opposition to Cape Wind, Kennedy's arguments do not hold up well to much scrutiny. Indeed, we were able to debunk most of them in just forty-five minutes of fact checking on Google and the Nexis news database.

In his *New York Times* piece, Kennedy claimed that the wind turbines would kill migrating birds. But the blades of today's wind turbines are so large and move so slowly that birds virtually have to run into them to be injured or killed. Studies of wind turbines in Europe similar to the ones that are planned for Cape Wind have found vanishingly few bird deaths annually.[2]

Kennedy pointed out that Cape Wind would bring in "40,000 gallons of potentially hazardous oil." And yet every year, hundreds of thousands of gallons of oil for the cape's electricity needs are transported across the sound in single-hulled tankers. Oil spills in the sound are no *potential* threat: in 2003, nearly 100,000 gallons of oil en route to a power station spilled into Buzzards Bay, killing 450 sea birds and contaminating beaches.[3] By contrast, the hazardous oil used by Cape Wind is *mineral* oil, which would be stored in stationary, triple-hulled containers.

Because of all the oil it burns for electricity, the Cape has some of the worst air pollution in Massachusetts. The group Clean Power Now, which supports the wind farm, used Harvard University's studies on power plants to conclude that Cape Wind alone would prevent twelve to fifteen deaths each year by reducing pollutants.[4]

Kennedy attempted to tap existing anxieties about the future of the New England fishing industry.

> Hundreds of fishermen work Horseshoe Shoal, where the Cape Wind project would be built, and make half their annual income from the catch. The risks that their gear will become fouled in the spider web of cables between the 130 towers will largely preclude fishing in the area, destroying family-owned businesses that enrich the palate, economy and culture of Cape Cod.

But most of Horseshoe Shoal is far too shallow for commercial fishing, and the cables Kennedy claimed would entangle fishing nets would be buried six feet under the ocean floor. In fact, the largest union of commercial fishermen — the International Seafarers Union — *endorsed* the wind farm.

Kennedy claimed the windmills would "be visible for up to 26 miles" — a statement that is patently false. Kennedy wrote that "lights to warn airplanes away from the turbines will steal the stars and nighttime views." But the lights would be barely visible — not nearly bright enough to have an impact on the stars. Kennedy claimed, "The noise of the turbines will be audible onshore." But visitors to similar wind farms in Europe reported that they could not hear the turbines even while sitting right next to them in boats.

In another op-ed for the *San Francisco Chronicle,* Kennedy claimed the turbines would pose "a dangerous navigational hazard to air and marine traffic."[5] In fact, the turbines will be spaced between six hundred and nine hundred yards apart, leaving ample room for ships to navigate. Kennedy claimed Cape Wind could be built farther offshore, but everyone else, including the Department of Energy, says such a possibility is ten to fifteen years away.

All of this, and yet over the following days and weeks there was no rebuke of Kennedy from national environmentalist leaders, accusing him of being insufficiently respectful of science, or of undermining efforts to win action on global warming, and certainly no reprimand from any of Robert Kennedy Jr.'s superiors at the NRDC.

It was left to Greenpeace and a small group of other activists who were sufficiently alarmed about global warming to speak out against Kennedy's distortions. In late December 2005, we joined Ross Gelbspan, author of *The Heat Is On* and *Boiling Point,* Bill McKibben, author of *The End of Nature,* and Jon Isham, a professor at Middlebury College, in drafting an open letter to Kennedy asking him to reverse his position. The letter was eventually signed by several hundred other grassroots activists.

2.

As lovers of beauty, we share Kennedy's appreciation of the nonhuman world. We believe this value must be brought forward from environmentalism and into the new politics. Nobody wants to live in an ugly world. But like the sailboats that glide across the sound, Cape Wind might actually *improve* the beauty of Nantucket.

For Bobby Kennedy, Cape Wind symbolizes precisely the kind of degradation of the Northeast's waterways that he has spent much of his career fighting, most famously along the Hudson. We share Kennedy's sadness and anger that so many of the Northeast's waterways and fish are contaminated. But that sadness and anger should motivate support *for* Cape Wind, not opposition to it. Not only won't Cape Wind poison the waterways, it will lessen the risk of future oil spills in Nantucket Sound, reduce the devastation to waterways in Ecuador and Nigeria from oil drilling, and advance the struggle against global warming, which threatens rivers, lakes, and oceans everywhere.

It is tempting to view the Cape Wind imbroglio as an anomaly. But beyond Kennedy's involvement, Cape Wind speaks to the problems that arise when one imagines that there is a thing called nature or the environment that is separate from and superior to humans, and that this "thing" is best represented by those who live nearest to it. It should go without saying that there are always conflicting priorities among local residents and visitors for the uses of a particular place. Indeed, Cape Wind's strongest defenders actually live in or around Cape Cod.

Believing in nature and defining one's politics around efforts to save it from human intrusion may have worked (at least temporarily) to protect places like the Arctic National Wildlife Refuge from oil drilling. The refuge comfortably fits the category of unspoiled nature; drilling comfortably fits as a destructive practice. Places like Love Canal worked well to demonstrate that certain human practices, such as dumping toxic waste where it can be breathed or swallowed, have destructive impacts upon people, not just on photogenic animals. But place-based environmentalism falls apart when faced with less easily categorized locales and practices, such as those in question in the case of Cape Wind.

Nantucket Sound is not Yosemite National Park. It has not been preserved in anything resembling a pristine, natural state. It has been a working human landscape for the better part of four hundred years. Even Yosemite, which we imagine as primeval nature in all of her glory, is itself a human construction. Everything from its nonhuman ecosystems (it has been regularly logged and ranched) to its protected status as a national park are the results of human intrusion.

Kennedy acknowledges that what is at stake is not so much nature as the human uses of it that have long been privileged on Nantucket Sound — sailing, diving, fishing, and bird watching — not generating electricity with windmills. But this is a clever reduction of the human uses in question. Nantucket Sound is, as Kennedy admits, among the busiest shipping channels in the Northeast. It is crisscrossed daily by tankers and ferries. It supports commercial and sport fishing, dive shops, marinas, and a variety of other industries.

The intrusion in question is no less complicated than the natural place Kennedy seeks to protect. Windmills are not oil rigs, which are by every estimation far more destructive, both locally in terms of their impact on the landscapes they sully and globally in terms of their contribution to global warming. Oil drills, whether in Alaska or the Ecuadorian Amazon, do more to threaten the Cape Cod landscape (natural or not) than Cape Wind or any other human activity in the area.

Perhaps the greatest irony is how often one hears the privileged residents of Hyannis Port lovingly describe those other, older windmills that dot Cape Cod as historic and even natural elements of the landscape. While some of the old windmills still grind corn, most of them no longer function in their original capacity — they are, for the most part, tourist attractions.

The problem with insisting that one is defending a singular and privileged nature from humanity is that the concept leaves environmentalists largely unable to distinguish among places like the Arctic National Wildlife Refuge, Nantucket Sound, and the couple of acres at the end of a cul-de-sac that nobody got around to developing twenty years ago. That's because the concept of nature in the theological, singular, and exterior way that environmentalists use it can only really be defined as "that which is not human," or, in the case of places environmentalists try to preserve, "that which has not been sullied by humankind." The result is that the Arctic, Nantucket Sound, and the end of the cul-de-sac all become roughly equivalent expressions of nature and, as such, equally deserving of human protection.

This is why Kennedy could compare Nantucket Sound to Yosemite and jeopardize the most important wind-energy project in the world while continuing to be lionized by environmentalists as one of their most charismatic and visible leaders. It is why national environmental organizations, except for Greenpeace, did virtually nothing publicly to back Cape Wind. And it is why there were no full-page ads, no celebrity conferences, and no flurry of e-mails and action alerts from the NRDC, Environmental Defense, or the Sierra Club calling on hundreds of thousands of their members to drop whatever they were doing and call Congress.

3.

An ethic born among the privileged patricians of a generation for whom building mansions by the sea was indistinguishable from advocating for the preservation of national parks or big-game hunting in the wilds of Africa turns out to be not very well suited either to grappling with the complexities of the new ecological crises or advocating ecologically responsible development in places like Cape Cod. In reducing the complexity of the nonhuman world to concepts of nature and place that are then depicted as essentially harmonious and unchanging, environmentalists often become obstacles to change and unwitting accomplices to the industrial status quo that they abhor.

In the case of Cape Wind, environmental opponents of the windmills end up functionally championing the continued dependence of Cape Cod and other Massachusetts communities on a nineteenth-century fuel source to heat their homes and generate electricity. Elsewhere, in the name of opposing development that is "out of scale to the neighborhood," they end up blocking the transformation of American communities into vibrant, creative, and high-density cities like New York that are far more sustainable and livable than endless megalopolises like Los Angeles.

Pick almost any NIMBY battle at random and you'll find similar dynamics. Whether it is Sierra Club members trying to kick mountain bikers off trails on which they like to hike, suburbanites fighting in-fill development in downtowns that are themselves only a couple of decades old, or national environmental figures championing immigration restrictions, NIMBY campaigns are almost always deeply conservative and almost always glorify places as they are in the present moment.

Landscapes, human and nonhuman, are always changing and evolving. When people describe something, whether it's new housing or a wind farm, as inappropriate to a particular place, what they are really saying is that it is inappropriate to a particular *idea* of that place. For instance, those who insist that dense housing development is inappropriate for their community or neighborhood often argue that dense development is out of scale with the existing community. Being out of scale with the neighborhood certainly seems like a reasonable — even ecological — objection. But when you think about it, increasing the density of any community by definition requires building structures, however beautiful or ugly, that are out of scale with what is already there.

Cities, and really all human communities, are as organic and *natural* as forests. In ancient cities like Rome and Paris, dwellings, churches, and streets, like an old-growth forest, are literally built upon the ruins of all that came before them. What is most compelling about those cities is not how they have been preserved but how they have evolved, how one can find ancient churches and landmarks cheek by jowl with modern, glass-enclosed apartment buildings. Are those modern buildings out of scale with what was there before? Well, yes. But so what? Being out of scale is only a problem if one imagines that the Rome of A.D. 100, or 1200, or 1700 is the essential or natural character of the place.

To embrace this notion is not to imagine that all development and change is necessarily good. It is rather to reject the idea that there is any singular, pure, or essential character to a place any more than there is a singular, pure, or essential nature. Windmills on Nantucket Sound or office buildings in Rome are not modern blemishes upon the essential character of those places. They are new human uses of landscapes that have served human needs and desires for centuries. New York is no more or less *natural* or *in scale* than Los Angeles or Washington, D.C., cities that restrict the height of their buildings. But New York, as a city and as a nature, works better for what we need right now: greater livability, efficiency, and sociability.

To contest the construction of in-fill development is to debate the idea of what kind of community one wishes to live in; to contest the construction of windmills on Nantucket Sound is to debate the idea of how, not if, Nantucket Sound will be used by people. And those questions are about which uses best serve our multiplicity of needs — for shelter, for community, for warmth and sustenance, for recreation — not about whether a particular place will be preserved in its essential state.

4.

The problem with having a conceptual framework that is unable to distinguish between good and bad development and that constantly focuses on preserving nonhuman natures, preventing development, and limiting growth is that when the time comes, environmentalists are ill prepared to support projects like Cape Wind or the development of dense cities as vigorously as they defend places like the Arctic refuge.

In opposing what they don't like, whether that's dirty energy plants, chemical agriculture, sprawl development, or interstate high-ways, environmentalists almost always argue that there is a better way — that wind energy, mass transit, and dense urban development offer not only better ecological outcomes but better human and economic outcomes.

Unfortunately, when push comes to shove, would-be green developers and entrepreneurs too often find themselves abandoned by the very same environmentalists who advocate such projects in theory. The idea that transforming America's economy, energy grid, or suburban landscape might require tradeoffs is anathema to many environmentalists, who routinely dismiss arguments that making the transition to a clean-energy economy might require putting up a few windmills in Nantucket Sound. To them, these are "false choices."

But the projected growth in carbon emissions and energy use over the next fifty years suggests that dealing with global warming will require us to make exactly the kinds of choices that environmentalists dismiss as false. China, for instance, will almost certainly bring the equivalent of the entire U.S. electrical grid online in the next twenty-five years. Coal is cheap and abundant in China and will be the fuel of choice there — unless cleaner, low-cost alternatives are readily available and easily scalable.

It is difficult to imagine an environmental movement that cannot even stomach windmills in Nantucket Sound engaging meaningfully in a politics that might help China meet its energy needs in ecologically sound ways. Such a politics would require environmentalists to accept and indeed embrace concepts like cleaner coal technologies and perhaps even nuclear energy. It will certainly require siting windmills in all sorts of places that people such as Bobby Kennedy Jr. might not like.

There is a difference between a *hard* choice and a *false* choice. An ethic that is based on preserving good natural things and stopping bad human things is poorly suited to either making hard choices or creating a politics capable of creating new good things. New identities, new landscapes, new economies, and new kinds of consumption will need to be developed to address the human and ecological crises of our time. Those who have a hard time accepting, nay, *embracing* windmills in Nantucket Sound or high-rise buildings in urban centers will not be able to create a politics capable of helping countries like China (not to mention the United States) meet their energy needs in a more ecologically sound way.

5.

Some will no doubt object that we are tarring all NIMBY efforts with the same brush and confusing authentic and positive efforts to protect nonhuman natures, such as watersheds, or to protect communities, like Love Canal, with self-interested efforts to preserve privilege and property. And indeed, local efforts to stop bad development and protect natural habitats are responsible for many great accomplishments. Consider that Kennedy's NIMBY alliance of wealthy, part-time homeowners has been effectively countered by a local Cape Cod organization, Clean Power Now, which boasts 6,500 members. Were it not for their commitment to a genuinely ecological politics, Cape Wind would have been killed long ago.

So it is not we who have conflated good and bad local activism, but rather environmentalists. Consider the way Kennedy embraced NIMBYism to defend his opposition to the wind farm.

> Yes, first of all, let me say this, *I believe in NIMBY.* I think it's the best part of American democracy. It's the essence of democracy. It's what saved the Hudson River. It's what cleaned up Love Canal. If you look at any of the major environmental victories we've had in this country it's because of people protecting their own backyard. If you think this is just about rich people you go talk to some of the fishermen down there and the marina owners . . . This is not just about wealthy people protecting their shoreline.[6]

The problem is not that all NIMBY campaigns are about wealthy people protecting their places, their views, their property values. It is that many of them are, and environmentalists either will not or cannot differentiate between good NIMBY efforts and bad NIMBY efforts. Environmentalists valorize NIMBYism in general as a kind of authentic and democratic place-based activism. Many believe, like Kennedy, that "it's the essence of democracy." And a big part of the allure of NIMBY politics is its simplicity: if everyone took care of his or her home, environmentalists tell themselves, the world would be a better place.

Grassroots environmental groups often accuse national environmental groups of being sellouts because they spend more time lobbying in Washington, D.C., than in organizing communities to protect beautiful places or fight pollution. But national environmental groups have long seen NIMBY campaigns as one weapon in their arsenal. Indeed, national environmental leaders see NIMBYism as key to their national power. In his response to our essay "The Death of Environmentalism," the Sierra Club's Carl Pope pointed to the power of place in winning action on global warming:

> Fully exploiting the potential of the pollution frame is, again, only one potential course for reframing the issue of global warming. Another is to take advantage of place-based environmentalism. One of the major global warming issues is that there are a huge number of coal fired power plants being proposed in the U.S. — about 112 gigawatts . . . Wherever they are built there are place-based advocacy tools to resist, which have been used quite successfully, say, in Colorado, as part of an integrated campaign to encourage wind and solar.[7]

But Pope himself knows that NIMBYism can be a double-edged sword. In our interview with him in 2004, Pope complained to us about the wealthy homeowners on Cape Cod who were opposing Cape Wind. "I was at the DNC [Democratic National Convention] and a pretty good person started haranguing me about windmills off Nantucket and how bad they are," he said before adding, "She didn't feel the slightest bit guilty in saying it."[8]

Despite Pope's support for Cape Wind, the Massachusetts chapter of the Sierra Club announced one year later that it would remain on the sidelines of the Cape Wind fight. "The Sierra Club has continued its position of being 'actively neutral,'" a member of the executive committee explained in the chapter's summer 2005 newsletter.[9]

Pope may have gnashed his teeth on hearing of the Massachusetts chapter's decision, but local opposition to Cape Wind was a case of the Sierra Club's NIMBY chickens coming home to roost. In the end, place-based campaigns are simply a neutral tactic, one that can be turned against wind farms as easily as against coal-fired power plants. Believing that there is something special and sacred about grassroots organizing makes no more sense than believing there is something particularly environmentalist about lawsuits, public relations, celebrity endorsements, full-page ads, or lobbying. Antienvironmentalists use these tactics to equal and often greater effect than environmentalists.

6.

We no longer believe it is justified to confine our affections to or reserve our loyalties for a particular race. Why then do we believe we are justified in reserving our loyalties for a particular place?

Ever since Darwin showed humans to be as much a part of the earth as a redwood tree or a hurricane, people have been forced to deal with the reality that our particular selves and circumstances — our race, place, moment in time, social class, aptitudes, families, and physical bodies — are accidents of chance. Given that, justifying one's politics according to one's *place* is as much a prejudice — and often as much a strategy to protect privilege — as justifying one's politics according to one's *race*.

And indeed, the fictions of race and place have long overlapped. People of a different race are initially from some other place. The environmentalist obsession with place-based politics is especially odd given how environmentalists emphasize the fact that we all share the same home: planet Earth. We may live in different places on earth, but we also live in a single place, where everything is connected to everything else. When you fully embrace this eco-logic of interconnection, a place-based politics starts to look as prejudicial as a race-based one. Why should it be acceptable for environmentalist residents of Cape Cod to declare themselves more loyal to Horseshoe Shoal than, say, to Ecuador's oil-poisoned rain forests?

The problem is not simply that it is difficult to answer the question "Who speaks for nature?" but rather that there is something profoundly wrong with the question itself. It rests on the premise that some people are better able to speak for nature, the environment, or a particular place than others. This assumption is profoundly authoritarian. The historic, worldwide transition to democracy means that decisions for how we use or preserve a particular place, whether that place be Cape Cod or planet Earth, will not and should not be handed off in advance to particular people because of their proximity to that place or their scientific expertise.

The NIMBY impulse is both conservative and undemocratic. It is conservative in that it is inherently averse to change. To imagine that the present state of a place or landscape is its natural state is to privilege the status quo through the invocation of a higher authority, namely nature. When people say they are being loyal to a place, what they imply is that said place has a single meaning and purpose outside of the meaning and purpose human beings give to it. What environmentalists are really being loyal to is a particular *idea* of a place — Cape Cod, for example, *without* a wind farm but with oil tankers.

Science tells us that almost all places — from the shifting continents to deserts that were once oceans to ancient cities that are literally built on the settlements that came before them — are in continual states of change and evolution, but environmentalists insist on privileging particular ideas as essential manifestations of particular places. They are thus asserting an unscientific way of seeing places — one that assumes stasis across time.

The idea that those closest to the land know best how to use, develop, preserve, or protect it is, ironically, invoked by both environmentalists and their opponents. In the eyes of many environmentalists, local resistance to environmental regulation and preservation of resources in areas like the American West is not place-based activism deserving of privilege but rather phony Astroturf activism conducted on behalf of corporations. At the same time, those property-rights and wise-use activists — the political labels often adopted by loggers, ranchers, and farmers who find themselves at odds with environmentalists — believe it is environmentalists, even those who live in their communities, who are outsiders attempting to impose their regulations at the expense of community well-being and prosperity.

Both sides attempt to privilege their viewpoints by asserting that their being closer to the land allows them to know the place better and hence make better decisions about its proper uses; they discount, marginalize, and demonize the views of others who are equally close to the land but hold opposing views.

To the extent that an individual justifies her politics *based on her vision for the place* — without claiming special authority because of her proximity to it — her politics are democratic. But to the extent she justifies her politics *based on knowing what nature wants* — or claiming special authority due to her proximity to the land — her politics are NIMBY and undemocratic.

7.

Environmentalists believe that their politics rest on "the facts," as discovered by science. But when we look at a windmill, a bridge, or a building, do we really see the same fact? Are the windmills ugly or beautiful? Are they symbols of our degradation of nature — or are they symbols of a brighter future?

"There's something attractive about harnessing the wind for energy," a Cape Cod resident observed on a Boston radio show. "I don't think it's going to take away from going to the beach. There's something good being done out there."[10]

The *Cape Cod Times,* which has long opposed Cape Wind, sent a reporter to Arklow, Ireland, south of Dublin, to see for himself what the planned tall wind turbines would look like. "About 6 miles offshore, the same distance from land as the proposed Nantucket Sound turbines, the seven turbines spin slowly on the Irish horizon at 15 revolutions-per-minute," Kevin Dennehy wrote. He interviewed local residents and discovered that the vast majority either liked or hardly noticed the turbines.

> "I can see them from my back window, but it doesn't bother me," said Eddie Breen, who owns a newsstand in the heart of Arklow, next to a Catholic church and across from a movie theater. "It's a good form of energy," Breen said. "Better than nuclear, that's for sure. If it breaks down, it's not going to emit any gases, is it?"[11]

Many observers have commented that they find the blades, which rotate languidly in even the fiercest gales, hypnotic. Tourists stare at them, reveling in their whisper-quiet power. Some people compare them to peaceful giants or scarecrows. But to our minds, the most delicious comparison is with those other wind-powered devices that grace the waters around Cape Cod: sailboats. From a distance of, say, six miles away, many people say that the white blades of a windmill stand against the blue sky much like the upright masts of Cape Cod's iconic vessels.

It is our bet that shortly after Cape Wind is completed, many of those who agitated against the windmills — perhaps even Bobby Kennedy himself — will smile and point to them with a local's pride.

5

THE POLLUTION PARADIGM

T OWARD THE BEGINNING of Al Gore's 2006 documentary
film *An Inconvenient Truth*, the former vice president shows
the first photograph of Earth taken from outer space. "The image exploded into the consciousness of humankind," Gore explains.
"In fact, within two years of this picture being taken, the modern environmental movement was born."[1] He quotes a NASA mission com-
mander reading from the Book of Genesis, "In the beginning God
created the Heavens and the Earth."

Gore proceeds to display photographs of traffic jams and power
plants belching visual pollution. Before we know it, we are seeing images of the apocalypse: hurricanes, floods, and droughts. "It was almost like a nature hike through the Book of Revelation," Gore says.[2]

Gore made *Truth* to warn us of our impending doom if we don't
limit our greenhouse gas emissions. It was promoted as "the most
terrifying film you will ever see." The tag line was "We're all on thin
ice."[3] And the media did its best to help. *Time* magazine published a
cover story promoting the film with an image of a polar bear literally
on thin ice, complete with the stark headline "Be Worried. Be Very
Worried."[4]

There was nothing in the movie or the accompanying book
aimed at helping viewers or readers imagine a brighter future for
themselves and their families. Gore could have dedicated the last half

of his slide show to describing an inspiring vision of the future, one centered on the creation of new jobs, new cities, and new profits out of new clean-energy industries. He could have spoken of revived economic competitiveness, international cooperation, and energy independence. He could have invoked America's can-do spirit with images of blue-collar workers building wind turbines in Ohio; suburbanites installing solar panels in California; and vibrant, clean, and livable cities in China. Instead, Gore spoke almost entirely of nightmares.

Toward the end of the slide show, Gore flashes a few familiar photographs of the fights for American independence, against slavery, against polio, for women's suffrage, against fascism, and for civil rights laws. But Gore's heart doesn't seem to be in it, and by the end he is back to showing pictures of planet Earth taken from outer space.

And that's the way the film ends. Apparently as an afterthought, the film's producers decided they needed to offer viewers something that they could actually do. So as the film credits roll, a list of "simple things" flashes on the screen: replace your light bulbs, insulate, don't consume so much stuff, and so forth. But the contrast between the problem and the solution alienated many environmentalists. The Pulitzer Prize–winning environmental writer Katherine Ellison wrote in an op-ed piece for the *New York Times:*

> Well, I for one am very, very worried. As the mother of two young boys, I want to do everything I can to protect their future. But I feel like a shnook buying fluorescent light bulbs — as Environmental Defense recommends — when at last count, China, India and the United States were building a total of 850 new coal-fired power plants.[5]

"The truth about the climate crisis," Gore explains, "is an inconvenient one that means we are going to have to change the way we live our lives."[6] At no point does one ever get the impression that this might be a change for the better. That's because, for Gore, it can't be. As surely as the Bible begins with a fall and ends with apocalypse, humankind's sins against nature will be punished. "It's human nature to take time," Gore says. "*But there will also be a day of reckoning.*"[7]

1.

Gore's movie was released on May 24, 2006, and generated more media coverage about global warming than the issue had ever received before, including cover articles in *Vanity Fair, Time,* and *Wired,* as well as extensive newspaper, television, radio, and movie reviews. Virtually every media outlet in the country, including Fox News, communicated Gore's terrifying truth about global warming.

The Pew Research Center for the People and the Press[8] conducted a telephone survey of 1,501 adults between June 14 and June 19, 2006, a period timed to coincide with the high point of the media's interest in Gore's movie. One of the central findings of the survey was that attitudes around global warming are highly partisan — a problem that Gore, who was widely reported to be considering another run at the White House, had likely exacerbated. Pew found that Democrats are more likely to believe the earth is warming than Republicans are (81 to 58 percent) and more likely to name humans as the cause of it (54 to 24 percent). Those are dramatic differences even at a time of great partisan polarization.

But by far the biggest finding was that the movie had done virtually nothing to increase the saliency of global warming among the public. Pew researchers noted that "out of a list of 19 issues, Republicans rank global warming 19th and Democrats and Independents rank it 13th."[9]

> While 41% say global warming is a very serious problem, 33% see it as somewhat serious and roughly a quarter (24%) think it is either not too serious or not a problem at all. Consequently, the is-

sue ranks as a relatively low public priority, well behind education, the economy, and the war in Iraq.

By January 2007, global warming's relative importance actually
declined to 21st out of 21 issues for Republicans, 17th out of 21 issues
for Democrats, and 19th out of 21 issues for independents.[10]

The goal of *An Inconvenient Truth* was to establish an overwhelming consensus for action based on dispassionate reasoning.
"Today, as the CO_2 crisis unites us," Gore explained, "we must remember the lesson of the CFC battle: that cool heads can prevail and
alter the course of environmental change for the better." But the
movie was fear-laden, not dispassionate, and rather than uniting us,
the environmentalist approach to the climate crisis has divided us,
which is why Pew researchers titled their report "Partisanship Drives
Opinion: Little Consensus on Global Warming."

Gore himself was careful to blame Republicans for his failure to
win action on global warming during his eight years as one of the
most powerful men in the world. "During the Clinton-Gore years we
accomplished a lot in terms of environmental issues," he insists, "even
though, with the hostile Republican Congress, we fell short of all that
was needed."[11]

But in the end, the problem was neither that global warming's
most visible messenger was a prominent Democrat possibly eyeing
the presidency nor that Gore had blamed Republicans for inaction,
but rather that the vice president and the rest of the environmental
community had, for more than twenty years, insisted that global
warming was essentially a problem of pollution to be fixed by a politics of limits.

2.

In 1947, a Harvard graduate student named Thomas Kuhn was working toward his doctorate in physics when he was asked by Harvard's
president to teach a class on the history of science. Kuhn, who was

twenty-five at the time, accepted the offer. He had never read any of the science classics before, and after reading Aristotle's *Physics,* Kuhn was struck by how different it was from Newton's theory of matter and motion. He concluded that Aristotle's science was not "bad Newton" but rather something fundamentally different from Newton. A few years later, he wrote up his argument in a popular monograph before seeing his *Structure of Scientific Revolutions* published in 1962, the same year Rachel Carson published *Silent Spring.*[12]

The phrase *paradigm shift* quickly became a buzzword among management gurus, New Age seekers, and environmentalists. And though the term is now part of our vocabulary, we often forget Kuhn's larger theory. This is too bad, because Kuhn's concept and his theory of how science actually works are vitally important to understanding not just scientific breakthroughs specifically but also social change generally.

Kuhn's argument was that science does not progress in a linear, rational way, but rather through "intellectually violent revolutions" where "one conceptual worldview is replaced by another." For example, Copernicus's theory was not a slight improvement over the Ptolemaic model; rather, it was a conceptual break. The same was true for Darwin's theory of evolution by natural selection, Einstein's theory of relativity, and Lavoisier's discovery of oxygen. These scientific revolutions established the new paradigms.

Revolutionary new paradigms — which include stories and theories about nature and reality, as well as cases, evidence, and facts — eventually become normal science, which proceeds over a period of decades to fill out the details of the once-revolutionary paradigm by assimilating new facts. A paradigm, whether revolutionary or established, both assists and restricts how scientists explain and predict the world and how they decide what the facts are.

Kuhn's book challenged the popular belief that scientists are skeptical, objective, and value-neutral thinkers. He argued instead that the vast majority of scientists are quite conservative; they've been indoctrinated with a set of core assumptions and apply their intelligence to solving problems within the existing paradigms. Scientists don't test whether or not they are operating within a good paradigm as much as the paradigm tests whether or not they are good scientists.[13]

Revolutionary science, Kuhn argued, often emerges when the established paradigm has an increasingly difficult time accounting for anomalies. Science is always dealing with anomalies, Kuhn pointed out, but when they become too prevalent or too persistent they can throw normal science into crisis. Faced with a growing crisis, the vast majority of scientists respond by jury-rigging the old paradigm to explain the anomalies. As the crisis deepens, the rules defining normal scientific practice loosen, and competing explanations of the crisis emerge. Often, the new paradigm emerges from *both* the crisis and the attempts to deal with it — eventually making the older paradigm irrelevant.

> Political revolutions are inaugurated by a growing sense, often restricted to a segment of the political community, that existing institutions have ceased adequately to meet the problems posed by an environment they have in part created. In much the same way, scientific revolutions are inaugurated by a growing sense, again often restricted to a narrow subdivision of the scientific community, that an existing paradigm has ceased to function adequately in the exploration of an aspect of nature to which that paradigm itself had previously led the way.[14]

What seemed to bother people most about Kuhn's theory was the notion that younger scientists embraced the new paradigm not for scientific reasons but because they had not yet been indoctrinated into the old paradigm. But this was something that the best scientists had understood long before Kuhn. The great physicist Max Planck had made a similar observation years earlier. "A new scientific truth does not triumph by convincing its opponents and making them see the light," he insisted, "but rather because its opponents eventually die and a new generation grows up that is familiar with it."[15]

One of Kuhn's most famous examples was of the revolution led first by Copernicus and later by Galileo to overthrow the Earth-centered view of the solar system and replace it with our current sun-centered one. But in other instances, new paradigms leave part of the old paradigms intact, such as Einstein's theory of relativity, which left Newton's theory of gravity *on Earth* intact even as it revolutionized our understanding of mass and energy in the rest of the universe.

Such may be the case with environmentalism. In many situations the pollution paradigm may still be a good way of understanding and dealing with air and water pollution. Our contention is not that the pollution paradigm is no longer useful for dealing with acid rain or rivers aflame but that it is profoundly inadequate for understanding and dealing with global warming and other ecological crises.

3.

In December 2004, Carl Pope of the Sierra Club wrote a 6,500-word response to our essay "The Death of Environmentalism," titled "There Is Something Different about Global Warming." Judging from the title, we thought that the reponse would grapple with the ways in which global warming fails to fit within the pollution paradigm. Instead, Pope insisted that global warming is essentially a pollution problem.

> Following this one line of possible alternative reasoning, how do we frame global warming as pollution? More particularly, how do we frame burning fossil fuel as pollution, because that is how the ordinary person will encounter this issue? Here's where it's not enough to think of global warming as a policy, or even a political problem. It's a conceptual problem. And it's a conceptual problem that environmentalism dealt with before, when it en-

countered the early view that "the smell of pollution is the smell of money."

Pope had in mind the public health fight against open sewers over one hundred years ago. "We won't make progress as long as we conceptualize fossil fuel consumption as a good thing (along the recent lines laid out by the World Bank) instead of presenting fossil fuel consumption as our century's open sewer."[16]

But Pope's open sewer example says more about the persistence of the pollution paradigm than it does about the alleged similarity between raw sewage and greenhouse gases. Whereas open sewers are visible for all to see, and stink to high heaven, the main greenhouse gas, carbon dioxide, is invisible and odorless. The only reason we consider carbon dioxide a pollutant at all is because we've produced *too much of it* — which is itself a challenge to the very meaning of "pollution."

Too much carbon dioxide in the atmosphere is, in this way, a classic Kuhnian anomaly. The "pollutant" itself occurs naturally. Unlike raw sewage, toxic chemicals, or carbon monoxide, it does not revolt us, poison us, or make us sick.

Having conceptualized global warming as essentially a problem of pollution, like open sewers in the nineteenth century, environmentalists brought to this qualitatively different ecological crisis the same paradigm that they had used to deal with smog, acid rain, and the depletion of the ozone layer. Phil Clapp, the head of the National Environmental Trust, one of the most influential environmental organizations in Washington, D.C., argued in his response to our essay that because a cap-and-trade system worked for acid rain it would work equally well for global warming.[17]

But dealing with the hole in the ozone layer was child's play compared with global warming, for it required a simple technical fix: the banning of chemicals suspected of depleting the ozone layer. Unlike coal and oil, those chemicals were never central to the basic functioning of the economy and had relatively simple and inexpensive substitutes. And dealing with acid rain meant not transitioning to entirely different sources of energy but fixing new pollution controls to a relatively small number of coal-fired power plants.

The great environmental laws of the 1960s and 1970s, while enormously important in protecting our air, land, and water, had the effect of shaping how all of us, not just environmental leaders, think about ecological crises. Those laws, from the Clean Air Act to the Clean Water Act to the many wilderness acts, were constructed around limiting human intrusions and contaminations of nature and the environment — things we unconsciously view as separate from humans. This is what we mean when we refer to the politics of limits. Environmental organizations whose mission it is to implement, defend, and extend those laws to cover new problems, from global warming to deforestation, thus borrow from the older mental models.

And herein lies the anomaly that most frustrates the environmentalists' pollution paradigm: the fact that overcoming global warming demands something qualitatively different from limiting our contamination of nature. It demands unleashing human power, creating a new economy, and remaking nature as we prepare for the future. And to accomplish all of that, the right models come not from raw sewage, acid rain, or the ozone hole but instead from the very thing environmentalists have long imagined to be the driver of pollution in the first place: economic development.

4.

When looking for signs of progress in the fight against global warming, environmentalists today frequently point to the European Union, whose members have all ratified the Kyoto treaty to limit greenhouse gases. The problem is that European nations have made little headway in actually reducing their own emissions. Germany and Britain have reduced their emissions, but most of those reductions were due to the collapse of the British coal-mining industry in the 1980s and the collapse of East German heavy industry and power generation after the reunification of Germany.[18] Little of the reduction in Britain or Germany is attributable to regulatory actions taken by either the European Union or national governments in the effort to reduce greenhouse gas emissions.

Greenhouse gas emissions throughout the rest of Europe and the rest of the developed world have either remained steady or increased. Japan will increase, rather than decrease, its emissions by 6 percent. Canada has seen a nearly 30 percent increase in emissions.[19] The Kyoto accords required signatories in the developed world to reduce their greenhouse gas emissions by 5 percent by 2012. What is becoming increasingly apparent is that few developed nations will meet their greenhouse gas reduction targets by reducing their own emissions.

Kyoto also allows signatories to meet their carbon reduction commitments by purchasing credits from other countries, and this is probably how most European nations will meet their targets. To date, they have done so by paying for carbon reductions in the *developing* world, through projects like retrofitting existing coal-burning power plants to burn more cleanly.[20] But as the Kyoto agreement's 2012 deadline approaches, and the developed nations of Europe struggle to meet their emissions limits, pressure from domestic industries will grow to allow them to purchase what is known as "Russian hot air" — the emissions credits given to Russia and the former Soviet-bloc nations for having reduced their emissions so dramatically in the 1990s. Like the collapse of British coal mining and East German manufacturing, the reason for the decline in greenhouse gas emissions in the former Communist countries had everything to do with the wider post-Communist economic collapse and nothing to do with Kyoto, global warming, or the environment.[21]

Whether the developed nations that are signatories to the Kyoto accords hit their targets by buying real emissions reductions or "hot air," there is little evidence to date to show that the Kyoto limits-based approach alone will succeed in actually reducing real emissions significantly in the developed world.

To have any substantial impact on the trajectory of climate change, most scientists agree that by the end of this century global greenhouse gas emissions will need to be reduced by roughly 70 percent of what they're currently projected to be if they continue to grow unconstrained. This will require much greater reductions in the *developed* world in order to allow for economic growth in the *developing* world. For instance, on a per capita basis, China emits less than 15 percent of what the United States emits.[22] What this means is that any viable long-term greenhouse gas emissions scheme must result in U.S. and Chinese emissions being stabilized at roughly equal levels, on a per capita basis, by late in this century. For that to happen, and to reduce emissions by 70 percent worldwide, U.S. emissions will need to decline not 70 percent from what its unconstrained emissions would be, but more like 90 percent from what they are today. Per capita emissions throughout the rest of the developed world are lower than U.S. levels, but will also still need similarly radical reductions.

Given the failure to date of European Community economies to make even modest reductions in emissions levels, it appears increasingly far-fetched that developed economies are likely to succeed in limiting their way to 80 or 90 percent reductions in greenhouse gas emissions. In the end, there will be no substitute for an investment-centered approach. Major public and private investments in the transition to clean-energy technologies will be necessary in order to achieve reductions on the scale that's required.

5.

If there are any lingering doubts about the failure of the pollution paradigm and the politics of limits to deal with global warming, one need only consider what's happening in China. The Kyoto approach, which imagined that growing emissions in China, India, and the rest of the developing world could be balanced by decreasing emissions in the developed world in the first decades of the twenty-first century, has been rendered irrelevant by China's great leap forward in emissions from cars and coal-burning energy plants.

Coal provides about 70 percent of China's energy, and China builds roughly one new coal-fired power plant every week.[23] It will pass the United States as the largest contributor of greenhouse gas emissions by 2008.[24] China's rapid economic expansion and corresponding increase in carbon emissions have so outpaced the projections of the Kyoto framework that this would be the case even if the developed nations of the world were making real progress on reducing their emissions, which they are not.

Global demand for energy, meanwhile, will grow 53 percent between 2006 and 2030, according to the International Energy Agency. Demand for coal will increase 59 percent, and oil consumption will increase from 85 to 116 million barrels a day. Carbon dioxide emissions, which are roughly 80 percent of greenhouse gases in general, will rise 55 percent, though some economists believe it will more than double unless we find cleaner energy sources.[25]

There is a scene in *An Inconvenient Truth* where Al Gore is greeted by cheering Chinese citizens and then seen conversing with energy experts as they huddle over a map of the nation. Gore and the filmmakers spend little time on the subject of China, perhaps because the Chinese have not shown a willingness to reduce their emissions unless they have some compelling economic reason to do so. The country rumbles with rising prosperity, rising expectations, rising demands for freedom — and rising discontent. The government manages environmental unrest the same way it manages labor and political unrest: through a combination of repression and cooptation. When there is some local environmental campaign or wildcat strike in a factory, the Chinese government responds with harsh punishments for leaders (sometimes even sending journalists who report on the protests to jail) and minor appeasements for workers and affected community members. Protest leaders are often imprisoned or sent to mental hospitals, while others receive the equivalent of a traditional red gift envelope: bite-size pay raises, pollution abatement, or personal freedoms calibrated to stave off discontent and allow China to continue its speedy climb toward greater prosperity.[26]

This is Beijing's strategy for dealing with everything from labor discontent to air and water pollution to demands for democracy. China, like Brazil, India, and every other developing country in the world, knows what its priorities are. "You cannot tell people who are struggling to earn enough to eat that they need to reduce their emissions," a Chinese government official told the *New York Times*.[27] And China knows where its contribution to global warming stands in comparison to the rest of the world. "It must be pointed out that climate change has been caused by the long-term historic emissions of developed countries and their high per capita emissions," said another government spokesman.[28]

The more coal and oil China burns, the more the government expresses concern over air pollution and global warming. Their concerns are not insincere. The problem is that environmental issues are not as high a priority as prosperity is. Educated Chinese will express their concern about global warming to you one minute before expressing their far greater excitement for their country's rapid economic growth the next. The government of China is indeed investing huge resources in efficiency and new energy sources, from wind to solar. But their motivations are simply that they need energy from any source they can get. They thus invest far more in coal and oil than anything else. Between 2007 and 2020, China will invest $128 billion in coal.[29]

On any given day, the country's English-language newspaper, the *China Daily*, is a testament to these contradictions. On December

27, 2006, an article headlined "Focus on Key Resources to Support Growth: New Large Oil and Gas Fields Found as Work Quickens on Energy Bases" sat directly above a much smaller article titled "Global Warming Could Wreak Havoc in China." A week earlier, the same paper had a lead editorial headlined "Shared Energy Challenge," which lamented rising oil prices, that ran next to a cartoon of a skier on a mountain with no snow, and above an editorial headlined "Global Warming" that could have been written by Al Gore.[30]

At the end of the day, the main goals of Chinese leaders are to stay in power and stave off unrest. The idea that the plaintive exhortations of Western environmentalists will persuade them to reduce their economic growth in order to reduce their emissions is implausible. A much better way to promote the transition to clean energy in China is to satisfy the public's desire and the government's need for rising security, stability, and affluence.

6.

On October 30, 2006, Nicholas Stern, the former chief economist of the World Bank and the British treasury secretary, released a seven-hundred-page report on the economics of climate change, which generated a thunderous reception in the world press. The report would become known as the "Stern Review," and the double entendre of its title was lost on no one. It declared that there would be massive economic consequences — "on a scale similar to those associated with the great wars and the economic depression of the first half of the 20th century" — if the world failed to act quickly to dramatically reduce greenhouse gas emissions.[31]

The report was significant for its scope and for the eminence of its author, as well as for the way it cast climate change centrally as an *economic* concern. But what was most impressive about the Stern Review was its holistic approach to action. It recommended that governments do three things at once: (1) tax carbon emissions, auction tradable permits to companies that emit greenhouse gas pollution, and reduce the number of permits over time; (2) dramatically increase investment and incentives for clean-energy research, development, and implementation; (3) prepare and adapt for the impacts of climate change.[32]

A few months later, in the spring of 2007, the United Nations International Panel on Climate Change (IPCC) issued new warnings of the seriousness of climate change and declared that 90 percent of global warming was caused by humans. And the IPCC proposed the same three-part strategy as the Stern Review: establish a price for carbon, invest in clean-energy alternatives, and prepare for a warmer and less stable world.[33]

Unfortunately, the environmental policy agenda has not yet caught up to the science of climate change. Between the release of the Stern Review and the release of the IPCC report, six major pieces of legislation to address global warming were introduced in the U.S. Senate. All were focused on establishing greenhouse gas emissions limits. None offered any significant new investments in clean energy or a serious plan to prepare for climate change.

Whether it's the recommendations presented by the IPCC, the Stern Review, *Scientific American*,[34] or top energy innovation experts,[35] investment is universally seen as a vital part of overcoming ecological crisis. "Funding for energy research," *Scientific American* said in its lead editorial in a special issue dedicated to clean energy, "must be accorded the privileged status usually reserved for health care and defense."[36]

There is no doubt that the effort to reduce and stabilize global greenhouse gas emissions will require a major regulatory effort to make sure that everyone is playing by the same rules, provide regulatory certainty so that nations and businesses alike can plan their investments, and increase the cost of fossil fuels relative to cleaner energy sources. But regulation cannot be the sole policy egg in the global warming basket, for without investments to encourage technological breakthroughs, we won't come close to achieving the emissions reductions we need to stabilize the climate.

7.

Kyoto epitomizes the environmentalist obsession with *limits*. The dominant proposals for dealing with global warming are thus conceptualized as strategies for regulation and sacrifice, not investment and opportunity. These proposals have been particularly unsuccessful in the United States, whose culture is infused with a spirit of inventiveness. But the environmentalist focus on limits also goes against the aspirations of Brazil, China, and India, countries that aim to become as prosperous as Europe, the United States, and Japan.

As we have seen, environmentalists have long defined their politics in the negative. They write books like *The Limits to Growth*, *The Population Bomb*, and *Collapse*, which focus our attention on the destructive power of humankind rather than on our creative potential. Environmentalists aim to *stop* logging, *restrict* overfishing, and *limit* pollution. They see in housing development only the loss of nonhuman habitat — not the construction of vital human habitat. Thus, the vast majority of environmental strategies aim to *constrain* rather than *unleash* human activity.

But in attempting to reimagine the ecological crises as opportunities, consider that limits have the potential to be creative. Restricting overfishing allows for the regeneration of fish stocks. Stopping logging allows a forest to grow. And limiting pollution, when done right, creates economic possibility. This is what happened with the Clean Air Act, the Clean Water Act, and the emissions trading system created to deal with acid rain. In each case, new technologies were invented, jobs created, and profits generated, all while achieving ecological objectives.

In the case of global warming, emissions trading — or what is known as "cap and trade" — could, if done right, generate billions of dollars in private investment for cleaner sources of energy. As such, it offers one of the best opportunities for environmentalism to evolve into a politics of possibility. Already, the European Union's emissions trading system is directing billions in investments to developing countries like China. Even though China and India have rejected limits to their own greenhouse gases, both are aggressively selling emissions reductions to European firms that need to reduce their emissions to meet European Union standards. So, for example, a European company that needs to cut its emissions can purchase emissions reductions from a Chinese factory, which uses the investment to become more energy efficient and thus emit less greenhouse gas. The regulatory mechanism created by the United Nations to oversee this process, through the Kyoto protocol, is called the Clean Development Mechanism, and it has already overseen the trading of $17 billion in emissions permits.

The result is that, at least initially, cap and trade will likely drive much greater investment in cleaner energy development in the developing world than in the developed world. And while major investment in the developing world is necessary to get the emissions reductions we will need, if it is not matched by a similar level of technology investment in the developed world, we will both fail to achieve the necessary emissions reductions and risk a serious political backlash, as voters in the United States and Europe realize that they are paying higher prices for energy, consumer goods, and services in order to pay for hundreds of billions of dollars in subsidies to improve the energy efficiency and independence of their economic competitors, China and India.

Keep in mind that the only vote that the U.S. Congress has ever held on the Kyoto plan was in 1997, when the Senate voted, 95–0, for a resolution against agreeing to any international approach that did not include comparable restrictions on China and other developing nations. Now imagine the attack ads that opponents of carbon caps might someday run: "Congressman Smith voted to increase your energy bill so he could send your job to China."

The only way Americans and Europeans will embrace such a scenario is if they perceive it to be a win-win arrangement. And for them to perceive it as a win-win, clean-energy investment, both public and private, must bring with it better jobs, economic growth, and new profits to the developed world as well as the developing world. To succeed over the long term, global emissions trading must be understood more as a national economic development agenda than as a regulatory framework to limit carbon emissions.

8.

In early 2007, there was a spate of breathless media reports announcing that clean energy would be to the early twenty-first century what computers and the Internet were to the late twentieth. The coverage described venture capitalist investments in everything from biofuels made from bioengineered algae to nanotech-based solar cells to next-generation ethanol as evidence that the transition to a clean-energy economy was already under way. What was largely missing from the hoopla was the recognition that none of the private sector investments in high-tech, computers, and the Internet would ever have occurred had the U.S. government not invested billions of the public's money in research and development and infrastructure in the fifties, sixties, and seventies.

The Internet (originally Arpanet) was created by the Defense Advanced Research Projects Agency, which was set up in response to the Soviet Union's launching of the first *Sputnik* satellite in 1957. In the 1970s, the Defense Department effectively guaranteed the market for microchips. The field of computer science would today be a marginal discipline had the federal government not spent billions on academic scholarships, fellowships, and other training programs, to lure the best and brightest young minds into the field. And it wasn't just computers. The invention of today's giant wind turbines was stimulated by incentives in the United States and Denmark in the seventies and eighties. The first solar photovoltaic cells were created for the U.S. space program in the 1950s.[37]

Big, long-term investments in new technologies are made only by governments and are almost always motivated largely by concerns about national security or economic competitiveness, from the threat of the Soviet Union in the 1950s to O P E C in the 1970s. Governments make the long-term investments in R&D and infrastructure, and the private sector capitalizes on them to develop specific products.

The problem is that similarly big, long-term investments are not being made in clean energy. The Stern Review notes that the roughly $34 billion invested each year in supporting clean-energy technological deployment (that amount includes nuclear) is "dwarfed by the existing subsidies for fossil fuels worldwide that are estimated at $150 billion to $250 billion each year."[38] It is for this reason that today, non-hydropower sources of clean energy — biomass, wind, geothermal, and solar — still represent only 2 *percent* of the world's electricity.[39]

While pushing for R&D investments in clean-energy technology might seem an obvious job for the environmental lobby in Washington, the truth is that pollution-oriented environmental groups have never prioritized those investments. As a result, public investment in energy research and development in the United States dropped from an already modest $8 billion in 1980 to $3 billion in 2005 (in 2002 dollars). In energy, as in computers and biotechnology, private investment capital follows public investment. Private venture capital during the same period dropped from $4 billion in 1980 to a paltry $1 billion in 2005. By way of comparison, a single biotechnology firm, Amgen, invested more than $2.3 billion in R&D in 2005.[40]

Part of the problem, the Stern Review notes, is that it takes decades before new energy sources become profitable. Private firms tend to be focused on generating returns in the short term — say, five to ten years — and not in the long run, twenty or thirty years in the future. New energy sources don't become profitable in small niche markets immediately, as many computer technologies do, because fewer people are willing to pay more for clean rather than dirty electricity.

It is frequently noted that there is enough wind, in places like the

Dakotas, to generate much of America's energy needs. But moving the electricity from windy and relatively unpopulated areas to densely populated cities, where most energy is consumed, will require billions in investments to upgrade our grid and allow for energy generation to be distributed. New clean energy depends on a new electricity grid, which as a public asset cannot be funded by the private sector.

The Stern Review recommends that governments boost clean-energy investments from current global levels of $34 billion to between $68 billion and $170 billion annually. Daniel Kammen, an energy analyst at the University of California, found that a domestic investment of $15 billion to $30 billion a year in clean-energy research and development is required to stabilize U.S. carbon emissions and create the market dynamics that would allow such investments to pay for themselves.[41]

What's needed is a portfolio of strategic, long-term investments. Government procurement of new technologies should be dramatically increased, public-private partnerships pursued, training programs created, prizes offered for technological breakthroughs, and international research collaboration encouraged. The Stern Review notes that it's not just research and development that needs funding, but also the creation of pilot projects, and the deployment of new technologies, ramping them up over time. With computer science and, more recently, biotechnology as the models, these public investments offer the promise of creating a vibrant new industry capable of driving economic growth for decades to come.[42]

9.

No single word better describes the ethics of environmentalism than "sacrifice." Ever since Henry David Thoreau spent twenty-six months living alone at Walden Pond, environmental virtue has been equated with a kind of self-denial. Forsaking excess consumption for the sake of the nonhuman world has been seen as the ultimate test of one's ecological commitment, and it was for this reason that through-out *Walden* Thoreau boasted of how little money he spent on food ($8.74) and materials ($28.12½) to build his spartan cabin.[43]

It's impossible to read *Walden* and not come away impressed by Thoreau's proto-environmentalist sincerity. But today, when relatively affluent environmentalists espouse sacrifice as the solution to global warming, their admonitions tend to ring hollow. When Senator John Kerry, for example, was asked in early 2007 what he had done personally for the environment, he answered that he and his wife "shifted to hybrid vehicles — not all of our cars, but I have a hybrid here in Washington, we have a hybrid in Massachusetts, we have a hybrid at our home in Idaho."[44]

Recall that the inconvenient truth that Al Gore named his book and movie after was neither that global warming is occurring nor that it is human caused, but rather that it demands sacrifice.[45] The day after *An Inconvenient Truth* won the Academy Award for the best documentary, Gore's hometown paper, the *Tennessean,* tipped off by a conservative think tank, revealed that Gore's home uses more than ten times as much energy as the average home in Nashville, and that his electricity bill is an astonishing $1,200 a month.[46] And that doesn't include the energy consumed, and the emissions produced, by Gore's other two homes.[47] Both Gore and Kerry say they purchase carbon offsets to make up for their emissions.

The problem is not that Gore and other prominent, wealthy environmentalists are failing to practice the sacrifice they preach. The problem is that none of us, whether we are wealthy environmental leaders or average Americans, are willing to significantly sacrifice our standard of living.

If purchasing carbon offsets and a hybrid vehicle for each of your three homes constitutes meaningful sacrifice, then perhaps we can reconcile the actions of environmentalists like Kerry and Gore with the rhetoric of environmentalism. But there is another problem: no serious energy economist or climate expert believes that individuals buying carbon offsets will play a key role in driving the transition to a clean-energy economy. We may achieve some greenhouse gas emission reductions by lowering our overall consumption, but the largest reductions will come from energy efficiency and shifting to cleaner energy sources — strategies that don't require drastic changes in the way we live our lives. What's needed, in short, is not so much *less* as *different* consumption.

Beyond being inadequate for reducing emissions, the politics of limits and sacrifice often leads environmentalists to arrive at some very dark conclusions. When the moment came for Gore to describe the causes of global warming in *An Inconvenient Truth,* he blamed the "population explosion" and flashed a photograph of logging in the Brazilian Amazon. "This rapid population rise drives demand for food, water, and energy — and for all our natural resources. It puts enormous pressure on vulnerable areas like forests — particularly the rain forests of the tropics."[48]

But Brazil has more than enough land and wealth to support all of its people without destroying the Amazon. Gore made no mention of the dictatorship debt, land inequity, or Brazil's reasonable desire for economic prosperity. And one can guess how Brazilians would feel about Gore's referring to *their* Amazon as "*our* natural resources." Gore noted that "most of the increase in population is in developing nations where most of the world's poverty is concentrated. And most of the increase is in cities." He presented this as part of the problem, yet cities, which are the most efficient places to house human populations, are precisely where those of us who are concerned about ecological crises should want people increasingly to live.[49]

The environmentalist obsession with "overpopulation" is typical of the view that ecological crises are essentially about too many people. But it has long been established that as families reach a certain level of education and wealth, their desire to have children goes down. It is for this reason that many European couples today are deciding against having children — and many European policymakers are beginning to worry about the social and economic consequences of *de*population.

The population reductionism of environmentalists represents a failure of imagination. Is it really so hard to imagine a world with healthy forests, a stable climate, and seven to ten billion people living in sustainable cities? How might history have turned out differently had we imagined the solution to global warming as *unleashing* rather than *restricting* human activity? What if we had conceptualized global warming not as the result of *too much* economic development but rather as the result of *too little* clean economic development?

10.

In recent years, advocates of sustainability have eloquently articulated a vision for a kind of natural capitalism.[50] What environmental leaders have so far refused to do is put this vision of human power, growth, and development at the center of their politics. Even when they establish national parks and wilderness areas, environmentalists see what they have done not as acts of *creation* but rather as *preservation*. They see themselves as *preventing nature's destruction* not *creating new kinds of nature*.

The anti-ecological logic of contemporary environmentalism reduces the cause of global warming to a single thing: humans emitting too much greenhouse gas. Their goal is thus to reduce greenhouse gas emissions. But what if we define the causes of global warming more expansively — as the consequence of our failure to create *new* economies, *new* patterns of development, *new* housing, and a *new* consumer culture, which together are far better able to meet our material and postmaterial needs?

Imagine what would have happened if environmentalists had proposed that the nations of the world make a massive, shared investment in clean energy, better and more efficient housing development, and more comfortable and efficient transportation systems. Opponents of global warming would have had to take the position *against the growth* of these new markets and industries and *for limits.* Proponents could have tarred their opponents as being anti-business, anti-growth, anti-investment, anti-jobs, and stuck in the past.

From one perspective, the situation today appears hopeless. If the developed nations most committed to action on global warming have made little progress in reducing their own emissions, what hope is there that China and India, whose energy consumption and emissions are growing exponentially, will be motivated to reduce theirs?

And yet, from another perspective, everything seems possible. There is quickly emerging a new political lobby and movement for clean-energy investment that is unburdened by either the pollution paradigm or the politics of limits. Increasingly, clean-energy companies and their private investors are realizing that they cannot rely on the environmental lobby and must take political matters into their own hands. And with young and grassroots environmentalists, more inspired by the vision of creating a new energy economy than regulating the old one, there's new hope that we will soon see the emergence of a more expansive, relevant, and powerful ecological movement.

In the debate over our call for the death of environmentalism, Bill McKibben, author of the 1989 *End of Nature,* reflected on the implication that the principal greenhouse gas pollutant, carbon dioxide, has no smell whatsoever.

> If you want to understand the death of environmentalism, you need to understand the gas on which it choked . . . Unlike carbon *monoxide* — the key ingredient in nasty brown smog, the pollutant that helped kill Londoners breathing coal fumes — carbon dioxide, ironically, is essentially nontoxic . . . Think about that, and perhaps you can understand why a political movement strong enough, barely, to protect blue whales and whooping cranes might be having a bit of trouble — and why any attempt to deal with

climate change will mean something that looks very different from environmentalism as we've known it.[51]

Humans are constantly rebuilding their civilizations around new visions and values. There is no reason they can't or won't do it again, to prepare for life in a warming world. The conventional wisdom on the global ecological crisis has it backward. Environmentalism is not the solution to the crisis of global warming. Instead, that crisis is driving environmentalism to evolve into something else. Reflecting on the birth of a politics capable of dealing with global warming, McKibben wrote, "If it has success, it won't be environmentalism anymore. It will be something much more important."[52]

6

The Death of
Environmentalism

R ACHEL CARSON OPENED *Silent Spring*, her 1962 polemic
against chemical pesticides in general and DDT in particular,
with a terrible prophecy: "Man has lost the capacity to foresee
and to forestall. He will end by destroying the earth."[1]

Silent Spring set the template for nearly a half century of environmental writing: wrap the latest scientific research about an ecological calamity in a tragic narrative that conjures nostalgia for Nature while prophesying ever worse disasters to come, unless human
societies repent for their sins against Nature and work for a return to
a harmonious relationship with the natural world.

Eco-tragedies are premised on the notion that humankind's survival depends on understanding that ecological crises are a consequence of human intrusions on Nature, and that humans must let go
of their consumer, religious, and ideological fantasies and recognize
where their true self-interest lies.

Part of the allure of the tragic narrative for environmental writers was that it *appeared* to be responsible for motivating action on the
pollution problems of the 1960s and 1970s. As we have seen, the less
obvious but far more powerful drivers for antipollution laws were
growing postmaterial desires, rising prosperity, and postwar optimism. In primarily crediting books like *Silent Spring* for the antipollution victories of the 1960s, environmentalists continue to preach

terrifying stories of eco-apocalypse, expecting them to result in the change we need.

And thus the new millennium brought with it a surge of environmentalist millenarianism. Grounded in a tradition of eco-tragedy begun by Carson and motivated by the lack of progress on the ecological crisis, environmental writers have produced a flood of high-profile books that take the tragic narrative of humankind's fall from Nature to new heights: Sir Martin Rees's 2003 *Our Final Hour*, Richard Posner's 2004 *Catastrophe*, Paul and Anne Ehrlich's 2004 *One with Nineveh*, James Kunstler's 2005 *The Long Emergency*, James Lovelock's 2006 *The Revenge of Gaia*, and Al Gore's 2006 *An Inconvenient Truth*, to name just a few.

For the most part, these environmentalist cautionary tales have had the opposite of their intended effect, provoking fatalism, conservatism, and survivalism among readers and the lay public, not the rational embrace of environmental policies. Constantly surprised and angered when people fail to behave as environmentalists would like them to, environmental writers complain that the public is irrational, in denial, or just plain foolish. Worse, they presume that the failure of the public to heed their warnings says something meaningful about human nature itself, attributing humanity's disregard for Nature to desires like the lust for power and concluding that, in the end, we are all little more than reactive apes, insufficiently evolved to take the long view and understand the complexity and interconnectedness of the natural systems on which we depend.

Kunstler begins *The Long Emergency* by quoting Carl Jung as saying, "People cannot stand too much reality." In fact, it was T. S. Eliot, not Jung, who said, "Humankind cannot bear very much reality."[2] But the attitude of such doomsayers recalls something Jung actually did say: "If one does not understand a person, one tends to regard him as a fool."[3]

Environmental tales of tragedy begin with Nature in harmony and almost always end in a quasi-authoritarian politics. Eco-tragic

narratives diagnose human desire, aspiration, and striving to overcome the constraints of our world as illnesses to be cured or sins to be
punished. They aim to short-circuit democratic values by establishing Nature as it is understood and interpreted by scientists as the ultimate authority that human societies must obey. And they insist that
humanity's future is a zero-sum proposition — that there is only so
much prosperity, material comfort, and modernity to go around. If
too many people desire such things, we will all be ruined. We, of
course, meaning those of us who have already achieved prosperity,
material comfort, and modernity. In the end, the story told by these
eco-tragedies is not that humankind cannot stand too much reality
but rather that Nature cannot stand too much humanity.

1.

Carson begins *Silent Spring* by narrating a "Fable for Tomorrow," describing a bucolic American town "where all life seemed to live in
harmony with its surroundings." The nearby farms flourished, the
foxes barked, and the birds sang in a kind of pastoral Eden. "Then a
strange blight crept over the area and everything began to change.
Some evil spell had settled on the community." Cattle died. Children
died. And the birds stopped singing. It was a silent spring.[4]

   Carson's fable, like most environmentalist accounts of Nature,
imagines Nature to be something essentially harmonious and in balance.[5] But long before there were humans, volcanoes erupted, asteroids hit Earth, and great extinctions occurred.[6] Fires burned through
ancient forests, killing undergrowth and creating the conditions for
new life. And throughout the animal kingdom there was murder and
gang rape — even among the much beloved and anthropomorphized
dolphins — activities that hardly qualify as harmonious. Indigenous
peoples, for their part, cleared forests, set massive fires, and over-
hunted, massively altering their environments. And they engaged in
agriculture, war, cannibalism, and torture.[7]

Why, then, do we continue to think of periods of calm as the earth's natural state and its frequent episodes of disruption and chaos as unnatural? Why is it natural when nonhuman natures (for example, asteroids) cause five mass extinctions but unnatural when human animals cause the sixth? If rape and murder are natural, does that make social prohibitions of them unnatural? And given that crude oil is natural, how could refining it and burning it be any less natural than, say, manufacturing solar panels from sand?

The categories of *Nature, the environment, natural,* and *unnatural* have long since been deconstructed. And yet they retain their mythic and debilitating hold over most environmentalists. Environmentalists imagine that they are objectively representing scientific facts about what is happening to Nature. But to imagine Nature as essentially harmonious is to ignore the obvious and overwhelming evidence of Nature's disharmony. To posit that human societies should model themselves after living systems that are characterized as Nature, as environmentalists so often do, begs the question: *which* living systems? Even if the earth heats up to such an extent that every last vestige of humankind disappears, there may still exist living systems, just not ones that can sustain us.

2.

In the Book of Genesis, the Fall from Eden occurs because Adam and Eve eat fruit from the Tree of Knowledge. In the environmentalist's telling of our fall, humans are being punished by Nature with ecological crises like global warming for our original sin of eating from the tree of knowledge — thus acting equal or superior to Nature. Our fall from Nature was triggered by our control of fire, the rise of agriculture, the birth of modern civilization, or sometimes, as in the case of *Silent Spring,* by modern science itself — which is ironic, given the privileged role the so-called natural sciences played in inventing the idea of a Nature as separate from humans in the first place.

The eco-tragedy narrative imagines humans as living in a fallen world where wildness no longer exists and a profound sadness pervades a dying earth. The unstated aspiration is to return to a time when humans lived in harmony with their surroundings. That tragic narrative is tied to an apocalyptic vision of the future — an uncanny parallel to humankind's Fall from Eden in the Book of Genesis and the end of the world in the final Book of Revelation.

Carson closes *Silent Spring* with these three grim sentences:

> The "control of nature" is a phrase conceived in arrogance, born of the Neanderthal age of biology and philosophy, when it was supposed that nature exists for the convenience of man. The concepts and practices of applied entomology for the most part date from the Stone Age of science. It is our alarming misfortune that so primitive a science has armed itself with the most modern and terrible weapons, and that in turning them against the insects it has also turned them against the earth.[8]

It is this reality — human agency — that most bothers environmentalists like Carson. For her, human attempts to control Nature inevitably end in tragedy.

In 1969, the microbiologist René Dubos won the Pulitzer Prize for a book calling for a new eco-religion based on the principle of harmony with nonhuman nature. "Whatever form this religion takes, it will have to be based on harmony with nature as well as man, instead of the drive to mastery."[9]

The contrast between living in harmony with Nature and mastering it is what unites Carson and Dubos with virtually every strain of contemporary environmentalism. Environmentalists imagine that their values are in opposition to the Western philosophical tradition, which sees humans as separate from and superior to Nature. This is what Carson speaks of when she refers to the "Neanderthal age of biology and philosophy." But what environmentalists challenge is the ordering of the categories, not the categories themselves. Rather than dissolving the distinction between humans and Nature, environmentalists reverse the hierarchy, arguing that humans are still separate from but *subordinate* to Nature.[10]

This reversal is motivated by the view that our perfectly healthy and natural desire to control our environment is a sinful desecration of Nature. But it must be asked: can human societies exist without, in one way or another, controlling Nature? Isn't that what agriculture is all about? Virtually any attempt to alter one's surroundings — whether by gathering wood to build a fire, constructing shelter, raising livestock, growing crops, or hunting and gathering — are efforts to control Nature. Nor is doing so uniquely human: beavers build dams, ants farm aphids, and more than a few other animals use tools.

There is nothing wrong with human and nonhuman desires for control over the environment. Indeed, we wouldn't exist were it not for our ancestors' will to control. Saving the redwoods and banning DDT were no less acts of controlling Nature than were logging ancient forests and spraying toxic pesticides. The issue is not whether humans *should* control Nature, for that is inevitable, but rather *how* humans should control natures — nonhuman and human.

From beginning to end, we humans are as terrestrial as the ground on which we walk. We are neither a cancer on, nor the stewards for, planet Earth. We are neither destined to go extinct nor destined to live in harmony. Rather, we are the first species to have any control whatsoever over how we evolve.

3.

The most sophisticated of the recent eco-tragedy books was Jared Diamond's 2005 *Collapse: How Societies Choose to Fail or Succeed,* a catalog of case studies of the deaths of past civilizations, such as the Mayaea and Anasazi, as well as contemporary societies, such as Rwanda during the genocide.

In *Collapse,* Diamond argues that past civilizations collapsed for five reasons: environmental damage, climate change, hostile neighbors, friendly trade partners, and societal responses to environmental problems. Diamond wrote *Collapse* believing that once people learned the facts of the ecological crises, they would make the rational choice to change direction. "I decided to devote most of my career efforts at this stage of my life to convincing people that our problems have to be taken seriously and won't go away otherwise," he wrote.

But in the end, *Collapse* is an argument against human attempts to control, ignore, or live out of balance with Nature. The stories that Diamond tells — of Greenlanders overgrazing their land and refusing to eat plentiful fish, of Easter Islanders and Maya kings deforesting their landscape in service of false idols — are tales of human hubris, of societies that neglected the laws of Nature in pursuit of human follies and were punished accordingly.

One of the primary cautionary tales in Diamond's book is of the Greenland Norse, whose civilization collapsed when irrational taboos triumphed over survival needs in a steadily worsening food crisis. The Norse colonizers of Greenland chose to starve to death rather than eat fish, "even though the Greenland Norse were descended from Norwegians and Icelanders who spent much time fishing and happily ate fish," and even though they saw the indigenous Inuit eating fish without negative consequence. Diamond wonders whether the Norse had developed a taboo against fish consumption, speculating that,

[P]erhaps Erik the Red, in the first years of the Greenland settlement, got an equally awful case of food poisoning from eating fish. On his recovery, he would have told everybody who would listen to him how bad fish is for you, and how we Greenlanders are a clean, proud people who would never stoop to the unhealthy habits of those desperate grubby ichthyophagous Icelanders and Norwegians . . . In trying to carry on as Christian farmers, the Greenland Norse in effect were deciding that they were prepared to die as Christian farmers rather than live as Inuit; they lost that gamble.[11]

Diamond argues that the Greenland Norse chose cultural survival — protecting their identities as beef and pork eaters — over biological survival. They knew what they needed to do to survive — eat fish. They simply chose not to do it.

The problem is that Diamond has no evidence for his claims, because the most important clues about the downfall of his collapsed societies — the values and beliefs and politics that led to their fall — simply don't exist. And the few facts they did leave behind tell us strikingly little. Diamond does not know why the Greenland Norse refused to eat fish, which is why he had to make up a story about Erik the Red and food poisoning.

Diamond is the author of major works on the human animal, the conquest of non-Europeans by Europeans, and the failure of human civilizations to adopt new values. But Diamond ignores several decades' worth of research into political psychology and social values, which offers far more clues to understanding today's ecological crises than the collapse of relatively tiny premodern societies.

Diamond gave *Collapse* the subtitle *How Societies Choose to Fail or Succeed,* in the belief that societies choose how and whether to adapt to changing circumstances. But neither the Easter Islanders nor the Greenland Norse ever convened tribal councils to choose collective suicide. The Easter Islanders, whom Diamond describes as having logged all of their trees in order to erect massive stone faces on their hillsides, and the Greenland Norse, who chose starvation over eating fish, were indeed behaving in ways that were perfectly *rational,* given their values, their cultures, and their belief systems.

In the end, Diamond projects the theological narrative of humanity's fall from Nature onto the societies that he writes about, asserting that "[as] a result of lust for power, Easter Island chiefs and Maya kings acted so as to accelerate deforestation rather than to prevent it."[12] In so doing, Diamond is unaware that he has told a *biblical* rather than a *scientific* story, a theological cautionary tale wrapped in the white laboratory coat of Science.

4.

For years, environmentalists have credited their strict and literal adherence to science for their successes, though not, notably, their failures. When environmentalism fails, it is invariably due to industry manipulation, the media's bias toward superficiality, the cowardice of politicians, public denial, and, most especially, the overall lack of deference in the United States to capital-S Science.

As we saw in chapter 1, environmentalists who came of age in the sixties grew to believe that the sweeping legislative victories of the era were the result of the public's literally seeing for themselves the destruction of Nature. But environmentalists also convinced themselves that it was their representation of Truth through Science that was responsible for antipollution and conservation laws. One prominent environmentalist-scientist perfectly articulated this view in an e-mail he sent to us and to national environmental leaders in the wake of the controversy over "The Death of Environmentalism."

> Mainstream national groups made a lot of progress over the past 40 years by maintaining scientific credibility as a touchstone of their efforts . . . Environmentalism identified with a scientific, rationalist approach, and drew a larger circle around itself and its positions accordingly.[13]

For many environmentalists, science is and should remain at the center of any effort aiming to overcome ecological crises. It is outside of history, society, and values. It is environmentalism's touchstone, the central criterion on which the value of environmentalism should be judged. But to believe that the sciences were behind the passage of environmental laws is a *faith* — a scientism, not a science — one that overlooks the specific historical and social conditions that gave rise to the ecological values.

The conventional wisdom is that environmentalists and global warming deniers like best-selling novelist Michael Crichton disagree over the value of science. But both share most of the same beliefs about Science and the need for it to stay clear of values and politics. This statement — "Because in the end, science offers us the only way out of politics. And if we allow science to become politicized, then we are lost"[14] — was uttered by Michael Crichton, but it could just as easily have been uttered by most environmental scientists.

Crichton's 2004 novel *State of Fear* is a formulaic thriller about environmentalist fanatics trying to force their values on society and the efforts of a single heroic scientist — the kind of rugged individualist Americans love — to expose them. What environmentalists misunderstood about Crichton is that he is motivated not by anti-environmentalism per se, but rather by scientific contrarianism. Consider that Crichton's 2002 novel *Prey* is a screed against nanorobots taking over the world, written in the same mold as an environmentalist cautionary tale such as *Silent Spring*. What is common to both *State of Fear* and *Prey* is a faith in the rugged scientific individual motivated purely by love of Truth.

Crichton ends his essay "Why Politicized Science Is Dangerous," the appendix to his *State of Fear*, with these words:

> The past history of human belief is a cautionary tale. We have killed thousands of our fellow human beings because we believed they had signed a contract with the devil, and had become witches. We still kill more than a thousand people each year for witchcraft. In my view, there is only one hope for humankind to emerge from what Carl Sagan called "the demon-haunted world" of our past. That hope is science.
>
> But as Alston Chase put it, "when the search for truth is confused with political advocacy, the pursuit of knowledge is reduced to the quest for power."

That is the danger we now face. And this is why the intermixing of science and politics is a bad combination, with a bad history. We must remember the history, and be certain that what we present to the world as knowledge is disinterested and honest.[15]

Crichton's antienvironmentalist narrative is structurally identical to the environmentalist narrative. His is a cautionary tale of what happens when one mixes a supposedly pure, apolitical, impartial Science with messy, corrupt, and partial politics. In Crichton's *State of Fear*, environmentalists willing to look can see their own scientism reflected back at them, distorted only modestly by Crichton's preference for the contrarian scientist.

5.

In premodern societies, when people claimed to speak for God, the discourse of religion and the power of the Church gave them a platform from which to do so. In modern societies, when people claim to speak for Nature, it is Science and environmentalism that provide the platform.

Environmentalists imagine that Nature, like God, is outside of us and all-powerful. And like God, Nature is telling us what to do. In 1989, when *Time* magazine named the Endangered Earth as Planet of the Year, its editors wrote, "This year the earth spoke, like God warning Noah of the deluge. Its message was loud and clear and suddenly people began to listen, to ponder what portents the message held."[16]

The environmentalist theologian Thomas Berry offered these words:

Now *the earth insists* that we accept greater responsibility, a responsibility commensurate with the greater knowledge communicated to us . . . *We need only to listen* to what we are being told through the very structure and functioning of our being.[17]

Berry insists that it is possible to be concerned with "the earth itself." He claims that he is "concerned with the earth not as the object of some human dream."[18] But that assumes the earth has a single dream, not multiple dreams, and that Berry knows what that dream is.

Berry claims that the earth's dreams can be experienced directly, without recourse to the sciences. But after you have read Berry's descriptions of the ecological crises we face, it is obvious that his understanding of what the earth is "dreaming" comes from the sciences. It is the sciences to which Berry refers when he asserts that "greater knowledge" brings with it "greater responsibility" for the earth's dream.

In this sense, Berry's environmental *theology* differs little from Carson's, Diamond's, or Dubos's environmental *Science*. The way environmentalists think of Nature is as metaphysical — and as authoritarian — as the way monotheists think of God. Like the fundamentalist who believes his religion only can speak for God, environmentalists believe that only Science, in a singular and objective way, can speak for Nature.

Just as there is no single transcendent Nature existing outside of humans, only differing ideas of what constitutes nonhuman and human natures, there is no single, glorious, and transcendent Science. There are only sciences creating contingent truths, toiling away to reveal, create, and organize facts and theories until the next revolutionary paradigm comes along to reorganize entire worlds.

Given that most of the intellectual founders of the environmental paradigm were scientists studying nonhuman nature, it's not surprising that environmentalism has constructed its politics so literally around objectively representing Nature through Science. Each of these scientists criticizes the way the sciences have been misused to destroy the environment. But few have doubted that Science should have a privileged role in shaping politics and human society.[19]

This faith in science is often accompanied by the antiquated view that there are facts separate from values and interpretations. But the fact that there is a strong international consensus among scientists that global warming is caused almost entirely by humans does not make it any less of an interpretation. And simply deciding *what* to study, and what kind of hypotheses to form, is a value judgment. The facts one chooses to give greater weight to in the case of global warming are deeply informed by one's values. The facts tell us that global temperatures have been rising over the last century. They tell us that human sources of pollution have probably been in some significant part responsible for those temperature increases. They tell us that global climate change and habitat destruction may be leading to the mass die-off of many plant and animal species.

But the facts also tell us that global temperatures have fluctuated wildly over the five billion years that the planet has existed; that there have been at least five previous mass extinctions during the history of the planet; that asteroids, comets, volcanoes, and ice ages have dramatically changed the climate and habitat at a planetary level; that the earth will very likely be here for billions of years after all traces of humanity have vanished from its surface; and that some form of humanity and human society will likely survive the ecological crises we face.

So which facts do we choose to focus on? Which conclusions do we draw? And what actions do we take based on those facts? These are questions as much about values as facts.

The questions before us are centrally about *how* we will survive, *who* will survive, and *how* we will live. These are questions that climatologists and other scientists can inform but not decide. For their important work, scientists deserve our gratitude, not special political authority.

What's needed today is a politics that seeks authority not from Nature or Science but from a compelling vision of the future that is appropriate for the world we live in and the crises we face. The idea that we should respect Nature implies that Nature has a particular single being (or dream) to be respected. If we define *Nature* as all things, then it is not at all clear *which natures* we should respect and which we should overcome. We are Nature and Nature is us. Nature can neither instruct our actions nor punish them. Whatever actions we choose to take or not to take in the name of the survival of the human species or human societies will be natural.

Many environmentalists imagine overcoming global warming to be about saving the planet. But the fate of the planet is not in question. The earth has survived meteorites and ice ages. It will certainly survive us.

6.

Carson, Diamond, Michael Oppenheimer, and others believe that Science and scientists, as the interpreters of Nature, can instruct our activities. The environmentalist and theologian Thomas Berry, as noted above, believes that the dream of the earth can instruct our political activity. But even Berry, despite his assertion that the earth's dream can be understood in an unmediated way, is almost wholly dependent on the sciences to describe it.

Given the status of scientists as the high priests of environmentalism, it should come as little surprise that it was a biologist and environmentalist, Edward O. Wilson, who invented the concept of *biophilia*, which he defines as humankind's "innate tendency to be attracted by other life forms and to affiliate with natural living systems" and which he believes is the key to the salvation of Nature and humankind.[20]

For Wilson, biophilia is something that can be experienced only in *nonhuman* natures, such as hiking through a forest, paddling a canyon, or sleeping in the desert. It is not something that can be experienced while making love, eating a meal with friends, or singing hymns in church. He refers to biophilia as an innate tendency — but what pleasure isn't? Is the pleasure we get from buying trinkets at the mall any less *innate* than the pleasure we get from walking through an ancient redwood forest?

To be sure, there are differences between sitting in a forest and sitting in a church. But what makes the former more *biological* than the latter? Are the mystical feelings of transcendence a Brazilian gets while sitting in a church in São Paulo any less natural — or powerfully felt — than those felt by Wilson walking through the Amazon?

Wilson implies that the feelings of awe and mystery we experience while in nature are more politically useful for saving nonhuman natures than those feelings we get when we are not in nature. But why? In his landmark essay on Nature, Science, and environmentalism, the political theorist Bill Chaloupka wrote, "Just because greens have become accustomed to the idea that Nature somehow instructs their political activity, it does not follow that there is no other way to motivate political activity."[21]

Couldn't the mystical feelings we experience while with other human animals be mobilized to support a politics capable of dealing with ecological crises? And why not biophobia, our all-too-natural fears of the weather? Couldn't we use that to mobilize support for action on global warming?

In his book *The Future of Life,* Wilson holds up as a beacon of biophilic politics Julia Butterfly Hill, a woman who, motivated by her biophilia, sat for two years in an ancient redwood tree, free from the messy reality of human society and politics. In the end, Wilson acknowledges, Hill saved only three acres. What he doesn't mention is that, tragically, soon after she descended, somebody skilled with a chain saw, likely an angry local logger, cut and nearly felled the tree.

Notwithstanding the political and cultural efficacy of those who, figuratively if not literally, make it their business to climb up trees and refuse to come down, what is most problematic about biophilia is the presumption that the tree told them to do it. But would we believe Nature was speaking to Hill any more than to the logger who cut down the tree?[22] What use is there in referring to what Nature wants, other than as a strategy to short-circuit democratic politics by asserting authority from a higher power?

Environmentalists such as Hill, Wilson, and Berry imagine that Nature is speaking to them through science and through the biophilic emotions that they feel when in a forest or other natural place. Place-based environmentalists imagine that they have special and privileged knowledge of a particular place based upon their proximity to it. Environmental justice leaders conflate conquest and oppression of nonwhites by Europeans and conquest and oppression of Nature by Europeans and then imagine that, as descendants of those who were conquered, victimized, and oppressed, they have special moral authority to undo the twin conquests of Nature and nonwhite peoples.

It is in this sense that we argue that all these different expressions of environmentalism are sectarian. They all claim special knowledge and moral authority, be it from Science, Nature, biophilia, place, or racial identity, that privileges their viewpoint and their politics. Sectarianism is, of course, ancient and hardly limited to environmentalists. Aristocrats claimed their authority came from God. Conservatives claim theirs comes from tradition. And fundamentalists claim theirs comes from faith and ancient texts.

But what all share is a perspective that claims to be above politics. Of course, all are partial and political. What the sectarian claim really asserts is that the particular viewpoint is inherently above all others. This is why environmental justice advocates constantly assert that any analysis of American history and politics that does not centrally feature race and racism is racist; why environmentalists constantly bemoan the politicization of Science; and why place-based environmentalists and wise-use antienvironmentalists alike demonize national environmental groups. In each case, the threat is that their claims to privileged knowledge and authority will be challenged.

7.

Frustrated by the public's unwillingness to defer to the higher aesthetic of Nature and Science, environmentalists increasingly shake their fists at what they imagine to be "human nature." In his book *Straw Dogs: Thoughts on Humans and Other Animals,* the British philosopher and environmentalist John Gray of the London School of Economics reduces the infinite complexity of human natures to self-destruction: "The mass of mankind is ruled not by its intermittent moral sensations, still less by self-interest, but by the needs of the moment. It seems fated to wreck the balance of life on Earth — and thereby to be the agent of its own destruction."[23]

For his part, the Sierra Club's Carl Pope blames our inner ape for environmentalism's political failures. "Global warming, habitat fragmentation, and the loading of global ecosystems with persistent but toxic and disruptive industrial chemicals are simply harder for an opportunistic, reactive primate species to understand as threats."[24]

Gray argues that we are simply ruled by "the needs of the moment," a claim that depends on complete and willful ignorance of more than fifty years of psychology and sociology. But even if this were the case, it is a statement so general as to be meaningless. *Which* moment rules us? Are we ever in the presence of a single moment? Aren't we always in the presence of multiple moments, past, present, and future? And don't all these moments have multiple meanings and possibilities? The contemporary moment no more calls on us to invade Iran and decimate the Amazon than it does to empower women and invest in clean energy.

Pope's generalization about humans begs related questions: If those who are concerned about the threat of global warming are members of the same primate species as those of us who are not, what explains the difference? If the failure to win action on global warming is because it *is* "intangible, global, and future oriented," in Pope's words, then wouldn't the logical remedy be to construct some alternative meaning for global warming, one that would inspire excitement, aspiration, and innovation?

Apes — whether human or nonhuman — are far more complex than either Pope or Gray acknowledges. Contrast the following account of our fellow apes offered by one of the world's leading primatologists, Dr. Frans de Waal, to the descriptions of the human animal offered by Pope and Gray.

> I call the human species the most bipolar ape, meaning that we go beyond chimps in our violence, which is systematic and often results in thousands of dead, and we go beyond the bonobo in our empathy and love for others, so that human altruism is truly remarkable . . .
>
> The main message from my book is that we should look at both of these close relatives, and consider that human nature is not all bad, nasty and selfish. It includes lots of positive tendencies. We are endowed with ways to hold aggression in check and to empathize, even with [our] enemies. These tendencies, too, are part of our evolutionary background.[25]

Gray wrote *Straw Dogs* not as a call for political action to overcome the crises but as a call for humans — whom Gray dubs *Homo rapiens* — to snap out of our denial and "simply see" that we have little control over our future.

> If anything about the present century is certain, it is that the power conferred on "humanity" by new technologies will be used to commit atrocious crimes against it . . . Nothing is more common than to lament that moral progress has failed to keep pace with scientific knowledge. If only we were more intelligent or more moral, we could use our technology only for benign ends. The fault is not in our tools, we say, but in ourselves.
>
> In one sense this is true. Technical progress leaves only one problem unsolved: the frailty of human nature. Unfortunately that problem is unsoluble.[26]

Humans are so destructive, Gray believes, that the earth will be better off when we are gone. "It is not of becoming the planet's wise stewards that Earth-lovers dream, but of a time when humans have ceased to matter."[27]

It's one thing for Gray to give up on the species. It is quite another for him to claim on behalf of Earth-lovers that humans have no way of controlling our behavior or shaping our future. It is true that we cannot excise our reactive, opportunistic, violent, and cruel natures. But neither can we excise our strategic, empathic, and altruistic ones. The "frailty of human nature" Gray describes is, from one perspective, insoluble. But in other contexts and from other perspectives, our frailties can be our strengths and our solutions. We do not need to solve the problem of our sins against Gaia, nor should we repent of our all-too-natural will to power; what we need to do now is to overcome destructive hatreds of who we are and antiquated prejudices about what we can become.

8.

Diamond's tragic narrative leads him to some disturbing political conclusions.

> People in the Third World aspire to First World living standards . . . Third World citizens are encouraged in that aspiration by First World and United Nations development agencies, which hold out to them the prospect of achieving their dream if they will only adopt the right policies, like balancing their budgets, investing in education and infrastructure, and so on. But no one at the U.N. or in First World governments is willing to acknowledge the dream's impossibility: the sustainability of a world in which the Third World's large population were to reach and maintain current First World living standards.[28]

Diamond believes he is being courageous in delivering this alleged truth.[29] But how courageous is it for Diamond to insist that the poorest people on earth should not aspire to the same standard of living that he himself enjoys? By the end of Collapse, Diamond has apparently terrified himself to the point of believing that there's simply not enough room on earth's lifeboat for everyone.

In chapter 1 we described how a common psychological effect of rising insecurity is for people to become more conservative, less generous, and more zero-sum: think pre-Hitler Germany or pre-genocide Rwanda. Many decades of social science literature strongly correlates rising insecurity, fear, and pessimism with authoritarian politics. In difficult situations, the insecure and the pessimistic seek out authoritarian leadership. What's more, social psychological research conducted in laboratory settings has found that manufacturing insecurity and fear, particularly of one's own death, can have the same impact as real social circumstances of fear, such as during a terrorist attack or rising economic insecurity.[30]

Affluent First World environmentalists like Jared Diamond are no less subject to the effect psychologists call "mortality salience" than are peasants in Rwanda. Little surprise, then, that after writing hundreds of pages of terrifying prose, scaring himself senseless, Diamond effectively embraces maintaining inequality and poverty in countries like China and India as well as the authoritarian view that humankind's survival rests on our ability to petition the rich and powerful. Here's Diamond's vision for political change.

> While few of us are personally acquainted with Bill Gates or George Bush, a surprising number of us discover that our own children's classmates and our friends include children, friends and relatives of influential people, who may be sensitive to how they are viewed by the children, friends, and relatives . . . The 2000 U.S. presidential election was actually decided by a single vote in the U.S. Supreme Court's 5-to-4 decision on the Florida vote challenge, but all nine Supreme Court justices had children, spouses, relatives, or friends who helped form their outlook.[31]

Five hundred and fifty-eight pages into Diamond's *Collapse* we discover that his political proposal is to petition Supreme Court members as though they were Anasazi high priests.

Collapse was intended to help Americans change their social values and create a more ecological society in order to avoid the fate of groups like the Greenland Norse. But in terrifying himself and his readers about the growing risk of social collapse, Diamond's eco-apocalypse narrative risks having the opposite effect. What extensive research finds is that the more scared people become about social instability and death, the *less* likely they are to change the way they think. "Fear of death," wrote a group of social scientists in 2003, "engenders a defense of one's cultural worldview."[32]

To direct our focus on collapse not only makes for a distorted view of human history, it risks undermining the security, confidence, and optimism required for progressive social and political change.

9.

There is a very different story that can be told about human history, one that embraces our agency, and that is the story of constant human *overcoming*. Whereas the tragic story imagines that humans have *fallen*, the narrative of overcoming imagines that we have *risen*.

Consider how much our ancestors — human and nonhuman — overcame for us to become what we are today. For beginners, they were prey. Given how quickly and efficiently humans are driving the extinction of nonhuman animal species, the notion that our ancestors were food seems preposterous. And yet, understanding that we evolved from being prey goes a long way toward understanding some of the feelings and motivations that drive us into suicidal wars and equally suicidal ecological collapses.

Against the happy accounts of harmonious premodern human societies at one with Nature, there is the reality that life was exceedingly short and difficult. Of course, life could also be wonderful and joyous. But it was hunger not obesity, oppression not depression, and violence not loneliness that were primary concerns.

Just as the past offers plenty of stories of humankind's failure, it also offers plenty of stories of human overcoming. Indeed, we can only speak of past collapses because we have survived them. There are billions more people on earth than there were when the tiny societies of the Anasazi in the American Southwest and the Norse in Greenland collapsed in the twelfth and fifteenth centuries, respectively.[33] That there are nearly seven billion of us alive today is a sign of our success, not failure.

Perhaps the most powerful indictment of environmentalism is that environmentalists so often consider our long life spans and large numbers terrible tragedies rather than extraordinary achievements. The narrative of overpopulation voiced almost entirely by some of the richest humans ever to roam the earth is utterly lacking in gratitude for the astonishing labors of our ancestors.

Of course, none of this is to say that human civilizations won't collapse again in the future. They almost certainly will. Indeed, some already are collapsing. But to focus on these collapses is to miss the larger picture of rising prosperity and longer life spans. Not only have we survived, *we've thrived.* Today more and more of us are "free at last" — free to say what we want to say, love whom we want to love, and live within a far larger universe of possibilities than any other generation of humans on earth.

At the very moment that we humans are close to overcoming hunger and ancient diseases like polio and malaria, we face ecological crises of our own making, ones that could trigger drought, hunger, and the resurgence of ancient diseases.

The narrative of overcoming helps us to imagine and thus create a brighter future. Human societies will continue to stumble. Many will fall. But we have overcome starvation, disease, deprivation, oppression, and war. We can overcome ecological crises.

10.

Ever since Darwin wrote *The Expression of Emotion in Man and Animals,* people have grappled with the ways that emotions are evolved, adaptive, and emergent. We have all of these wonderful emotions within us for good evolutionary reasons. We should be grateful to them all, for they have helped us thrive. We must now learn how to make new use of them as we head into the future. Embracing a pragmatic, ecological, and scientific multinaturalism demands that we let go of the outmoded idea of the singular, natural, and essential self.[34] We are a welter of genes, ideas, chemicals, mental organs, instincts, emotions, beliefs, and potentials colliding inside and outside of our skin.

There are good evolutionary reasons for cruelty, hatred, and violence, as well as for altruism, empathy, and cooperation. Our challenge is not simply to overcome our worst natures or to strengthen our best ones. It is to recognize that some natures, like some values, are appropriate for some stages of human development but not others.

There is no single spirit or essence that defines us. Humans are not essentially opportunistic, reactive, conservative, adaptive, creative, or destructive. We are all of the above and more. In embracing the fact that we contain multiple natures, we abandon the notion that humans are *essentially* anything. Putting any particular nature, rationality, or instinct at the center of our being strangles our potential.

A new politics capable of dealing with the great ecological and other human crises we face must embrace the scientific exploration of human natures and not just the scientific exploration of nonhuman natures, like global warming. And we must start seeing humans as highly adaptive creatures capable of great acts of destruction and creation.

The philosopher Richard Rorty has suggested that we borrow from the Romantics the view that in politics, as in life, imagination is more important than reason.[35] If he is right, then environmentalists have gotten their politics backward. For too long they have demanded that Americans *Wake up!* rather than encouraging them to dream. But Rorty also recognizes that, in the end, all imagination entails a kind of rationality, and all rationality rests on a kind of imagination.

The question is not "Reason or imagination?" but "*Which* reason? *Which* imagination?"

How might history have turned out if environmentalists had adopted a generous, adaptive, contingent, and *anthrophilic* politics that flowed from an affirmative and expansive vision of our natures? How might it have differed had environmentalists adopted an egalitarian vision that embraced politics, in all its untidiness and contestations, rather than simply substituting the new authorities of Science, Nature, place, and race for the old authorities of God, Man, Progress, and Tradition?

A new politics requires a new mood, one appropriate for the world we hope to create. It should be a mood of gratitude, joy, and pride, not sadness, fear, and regret. A politics of overcoming will trigger feelings of joy rather than sadness, control rather than fatalism, and gratitude rather than resentment. If we are grateful to be alive, then we must also be grateful that our ancestors overcame. It is thanks to them, and the world they made, that we live.

The political theorist Jane Bennett suggests in *The Enchantment of Modern Life*, "This life provokes moments of joy, and that joy can propel ethics." Bennett's book is a happy deconstruction of the belief that the modern life objectifies and disenchants the world, robbing it of its mystery, ineffability, magic, and connectedness. Bennett insists that the world never lost its capacity to surprise and inspire. She argues for an ethics that begins with a commitment to affirming life in all of its joys and sufferings.

> If popular psychological wisdom has it that you have to love yourself before you can love another, my story suggests that you have to love life before you can care about anything. The wager is that, to some small but irreducible extent, one must be enamored with existence and occasionally even enchanted in the face of it in order to be capable of donating some of one's scarce mortal resources to the service of others.[36]

The ethics, and politics, born of the joy, mystery, and gratitude of overcoming adversity will be radically different from the ethics born of the sadness of living in a fallen world pervaded by fears of the eco-apocalypse to come. The truth is, there are still ancient redwoods to behold and great rivers to swim in. There is still the Amazon and the Boreal. There are still seven billion wondrous human animals, each one of us capable of making ourselves into something utterly unique. And there is still great wildness abounding inside and outside of ourselves.

PART TWO

THE POLITICS OF POSSIBILITY

We have it in our power to begin the world over again. A situation, similar to the present, hath not happened since the days of Noah until now. The birth-day of a new world is at hand.

— THOMAS PAINE, *Common Sense*

7

STATUS AND SECURITY

W$_{\text{HAT'S THE MATTER WITH KANSAS?}}$, Tom Frank's 2004 jeremiad against the Republican coalition of Christian conservatives and market fundamentalists who had come to dominate politics in his home state, opens with an epigraph: "Oh, Kansas fools! Poor Kansas fools! The banker makes of you a tool."[1] The line, which is from an 1892 populist song, neatly sums up Frank's view, shared by many liberals and Democrats, that the success of the modern conservative movement was the result of Republican success in manipulating working-class Americans into voting against their material self-interest.

Kansas is, in this way, an ancient narrative about the deployment of bread and circuses by the ruling class, wrapped in hip, bristling prose. The gag that runs throughout the book has Frank constantly turning to the readers as though to say, *Can you believe how deranged my people are?*

Like environmentalists who see their fellow humans as essentially opportunistic and reactive, Frank tends to view the average conservative Kansan as someone whose ability to reason dispassionately has been occluded by the passions.

> The trick never ages; the illusion never wears off. Vote to stop abortion; receive a rollback in capital gains taxes. Vote to make

our country strong again; receive deindustrialization. Vote to screw those politically correct college professors; receive electricity deregulation. Vote to get government off our backs; receive conglomeration and monopoly everywhere from media to meatpacking.[2]

What liberals like Frank and environmentalists like Jared Diamond have in common is the view that the primary obstacle to progressive political action is that Americans are ignorant and easily manipulated. Frank describes the Great Backlash in largely the same way that Marxists used to describe false consciousness. "People getting their fundamental interests wrong," he writes, "is what American political life is all about."[3]

Frank powerfully documents the success of conservatives in winning the hearts and minds of working-class Kansans. But he misdiagnoses the reasons for the rise of conservative power. Frank claims conservatives succeeded by diverting voters' attention away from their material needs. What Frank overlooks is the fact that Kansans more than met their fundamental interests long ago. The reason most Kansans are no longer attracted to the old New Deal agenda is that most Kansans are no longer poor.

1.

In 2004, when Frank's book was published, the annual median family income in Kansas was $43,725, compared to $44,473 nationwide.[4] In other areas, Kansas is doing better than the rest of the country. For example, both its unemployment and poverty rates are lower.[5] Moreover, Kansans live longer,[6] smoke less,[7] are more likely to own homes,[8] have health insurance,[9] graduate from high school[10] and college,[11] and have shorter commutes[12] than the average American.

Frank asserts, "The more working-class an area is, the more likely it is to be conservative. This situation is the opposite of what it was thirty years ago. And it is the complete and utter negation of the Kansas of a hundred years ago."[13] But thirty years ago and a hundred years ago, the average Kansan was far more conservative around issues like marriage, abortion, homosexuality, gun safety, and race relations. Indeed, the reactionary values of Kansas backlashers are not new — they're *traditional*. When Frank praises working-class Kansan populists of the past for being radicals, he uses the term in the old left sense: they wanted higher taxes on the rich, more help for the poor, and greater government regulation of corporations.

Frank's heroes, the populists of the turn of the century, were far from being compassionate, generous progressives. They were insecure, desperate, and often quite mean and prejudiced. As Harvard economist Benjamin Friedman noted in his recent examination of the connection between economic growth and social values, "The populists of the late nineteenth century felt, all too concretely, the pain of being dispossessed, both economically and culturally. They saw conspiracies among bankers, foreigners, Jews, railroads, and anyone not belonging to their own part of America, and especially anyone who appeared to be succeeding as their circumstances deteriorated."[14]

What's new is not the moral values of working-class Kansans but rather the rising political import of those values against a backdrop of materialist liberalism in decline. As Kansans became more affluent after World War II, they quite understandably felt they needed government less. Meanwhile, the liberalizing of public attitudes and behaviors around marriage, sexuality, and abortion sparked a traditionalist backlash, and evangelical conservatives made a shrewd alliance, ideological and otherwise, with libertarian conservatives. Together they channeled their resentments into a politics powerful enough to take over the Republican Party and push Kansas to the right. The problem is not that Kansans have not met their basic survival needs, as Frank would have it, but rather that those religious conservatives seeking things like restrictions on abortion and banning gay marriage have better harnessed postmaterialist desires for status and belonging than progressives seeking to promote things like universal health care.[15]

But Frank is also onto something very important, and that's the slow rise of insecurity since the mid-1970s and the gradual return to what we have been calling survival values, such as xenophobia, patriarchy, and the acceptance of violence. But unlike during the Great Depression and New Deal, the insecurity felt by reactionary Kansans is *postmaterial*, not material — what we will describe throughout this chapter as insecure affluence.

Frank points to Kansas's slippage since the early seventies but ignores its astounding prosperity during the thirty prior years. Kansans today struggle not with hunger but with obesity. To the extent that a working-class Kansan is insecure it is because he fears losing his quality of life — not his life.

2.

In part I of this book, we attempted to demonstrate the ways in which environmentalism has failed to recognize or concern itself with the social preconditions that gave rise to it and has instead fallen back on an ancient narrative of limits and a modern one of rationality and science. In part II, we'll address the larger social context and conditions necessary to create progressive social change. And we will describe the broad outlines of a new politics that we believe can take us where we need to go.

That politics will be postenvironmental, not environmental, and postmaterial, not material. It will be an exercise not in speaking truth to power but in creating new truths in the polity. And it will understand political identity and behavior not simply as reflections of ideas rationally arrived at but also as expressions of the social and psychological needs each of us experiences in both predictable and unique ways.

We began this book by reflecting on the conditions that gave birth to environmentalism and other progressive movements for social change of the 1960s. We noted that Americans were materially poorer then than they are today, but they also had higher expectations for a brighter future. We referred to the very strong sociological evidence from around the world showing that this combination of rising prosperity and optimism almost always leads to a move away from traditional forms of authority and toward greater individual autonomy, and a stronger embrace of self-expression and what we have been calling fulfillment values.

We tend to look back on the sixties through lenses darkened by our memories of the Vietnam War and the urban riots at the end of the decade. But in the late fifties and much of the sixties, Americans felt proud of their victory over fascism in Europe and Japan and their success at turning former enemies into prosperous, democratic allies. They were optimistic that the green revolution might eliminate hunger and that economic development and technological innovation might create both prosperity and greater leisure for all. This pride and optimism, combined with prosperity and the strong and widespread emergence of postmaterial fulfillment values, led to the end of legal racial segregation, discrimination, the expansion of the social safety net, and efforts to clean up our air and water.

But today, America no longer sees itself or the world through the rose-colored glasses of postwar optimism. The postwar economic expansion, which saw virtually every American household experience consistent increases in the standard of living, came to a halt in the early 1970s and stagnated for the better part of twenty years. Optimistic visions of a peaceful and prosperous world led by a powerful, generous, and idealistic superpower gave way to dystopian visions of ecological collapse and overpopulation. And while the Vietnam War undermined faith in the efficacy and morality of American military power, the Watergate scandal undermined trust in American government.

If the rising tide of prosperity and optimism floated all boats in the postwar era, engendering an expansive and generous politics, the darker mindset of recent decades has led to a zero-sum, conspiratorial, and fear-based politics on both the political left and right. Environmentalists assert that there is not enough prosperity to go around and warn of collapse and apocalypse. Much of organized labor, in contrast to its midcentury internationalism, opposes global trade in a desperate effort to hold on to what remains of the U.S. manufacturing economy. Much of the left opposes virtually all expressions of American power in the world. Liberals, even as they bemoan the decline of American faith in government, see government as profoundly broken and corrupted.

America's evolving social values underlie and reflect these changes. Americans have increasingly moved away from higher-order to lower-order postmaterialist social values — from inner-directed fulfillment values toward outer-directed materialistic ones. Rising prosperity has resulted in an increasing orientation toward personal freedom and choice. But these values, combined with greater affluence *and* declining expectations and optimism, have resulted in rising materialism, outer-directed status-oriented behaviors, and fatalistic survival values.

In order to work our way toward a new politics, one that is both pragmatic and inspiring, we need to take a closer look at America's move back down Maslow's hierarchy of needs, from higher-order fulfillment to lower-order status. We must grapple with the ways in which Americans have become both more affluent and more insecure. And we must come to terms with the ways that liberalism, which was born of the material scarcity of the Great Depression, fails to speak to postmaterial needs, born of American abundance.

3.

Just as one cannot appreciate the birth of environmentalism without understanding the material prosperity from which it emerged, one cannot comprehend modern liberalism without understanding the material scarcity from which it was born. In *A Paradox of Plenty*, Harvey Levenstein describes the hunger of the Great Depression.

> Within a few months [into 1930] there are some eighty-two breadlines operating in New York City — two more than in Philadelphia, the nation's third-largest city. Halfway across the continent, food riots break out in small towns called Henryetta, Oklahoma, and England, Arkansas, as hungry crowds shouting "We want food" and "We will not let our children starve" threaten local relief agencies and merchants . . .
>
> In 1931 Americans were shocked to read of people digging for rotten food in garbage dumps in St. Louis. In Harlem men and women could be seen competing with dogs and cats for the contents of garbage cans . . . Kentucky hill people were reported to be living on dandelions and blackberries. Conditions in the Appalachian coalfields were said to be particularly grim. The miner's diet often consisted only of beans ("miner's strawberries") and "bulldog gravy" (flour, water and grease) for his lunch pail.[16]

President Franklin Delano Roosevelt was elected in 1932 and soon after announced the New Deal. But the Great Depression would not end until the United States entered World War II, in 1941. In 1939, 9.4 million American workers were still without employment. By 1944, with the United States three years into the war, that number had declined more than 90 percent, to just 670,000.

In 1946, the year after Americans had helped vanquish fascism and overcome the Great Depression, they were arguably the happiest they have ever been. A powerful feeling of social solidarity pervaded American society. The affluence that the United States proceeded to achieve was nothing short of extraordinary. Between 1948 and 1973, the median family income grew about 3 percent each year, going from $20,600 to $43,200. Unemployment hovered under 5 percent, and economic growth averaged a whopping 3.8 percent annually.[17] Medical advances, from the discovery of a polio vaccine to the widespread availability of antibiotics, saved lives and extended life expectancy from forty-seven in 1900 to seventy-seven today.[18] The rising tide not only lifted almost all boats, it lifted them in ways that were widely perceived as fair, thus creating a sense of collective aspiration and optimism.[19]

That prosperity and optimism gave rise to the civil rights, antiwar, student, and women's movements of the 1950s and 1960s. Rather than being countercultural, these political movements were consistent with the rapid climb Americans were making up Maslow's hierarchy of needs.[20] Black Americans, who had long been excluded from the economy and society, were finally wealthy enough to demand their higher needs for their political and civil rights, and white Americans were finally secure enough to grant and even fight for them.[21] For the relatively privileged whites who supported the civil rights movement, rising prosperity meant overcoming their *outer-directed* status needs and seeking to fulfill their *inner-directed* needs for purpose and meaning.

The baby-boom generation's dramatic move away from traditional forms of authority — the patriarchal family, militaristic government, authoritarian schooling, and corporate hierarchy — in search of authenticity, spirituality, personal creativity, and a connection to the nonhuman world was made possible by the fact that those young people had not only met their basic material needs but had overcome the first level of postmaterialist desires for self-esteem, status, and recognition.

By the time President Lyndon Johnson declared a war on poverty in 1964, America was well on its way to becoming what sociologists call a postscarcity society. In 1964 and 1965 the government established Medicare and Medicaid, created a food-stamp program, extended greater assistance to schools in poor communities, and expanded job training. In the late sixties, President Nixon proposed laws giving automatic cost-of-living adjustments for Social Security and expanding welfare. He even fought for a negative income tax, which effectively gave government money to poor working families, and which was eventually passed by Congress as the Earned Income Tax Credit and signed by President Ford.[22] By 1970 Americans were so well-off that President Nixon could give a State of the Union address calling on Americans to put happiness and "peace with nature" before greater material wealth.

4.

Whereas the median family income grew an astonishing *100 percent* between 1948 and 1973, it grew just *7 percent* between 1973 and 1993 — this despite the fact that huge numbers of women were entering the work force, creating a record number of two-income families. By 1990, a higher percentage of the work force was employed than during World War II.

The reason the median family income increased just 7 percent despite so many more two-income families was that the average real wages for nonmanagement workers had declined dramatically, from $631 per week in 1973 to $491 in 1993 (in 2004 dollars).[23] Young people were particularly affected. A thirty-year-old man earning the median wage in 1973 earned *two-thirds* more than what a thirty-year-old man had earned in 1953, while a thirty-year-old man in 1993 earned a quarter *less* than what a thirty-year-old man had earned in 1973.[24]

Stagnating wages were exacerbated by rising inequality, and this in turn fed increasing status insecurity. Between 1979 and 1989, the richest 1 percent of American households increased their income from $280,000 to $525,000 per year.[25] It wasn't just that people were working harder to earn less, but that they were doing so at a time when other people's material success was being widely publicized in hit TV shows like *Lifestyles of the Rich and Famous*. The seeds of insecure affluence were planted.

In his 2005 book *The Moral Consequences of Economic Growth*, Benjamin Friedman assembled an impressive body of evidence showing how, during times of economic growth, people become more empathic, expansive, and generous toward others, including immigrants, racial minorities, gays, and women. And he described how we become less generous and more status conscious during times of rising insecurity, such as the period after 1973.

> The impact of this economic stagnation on Americans' attitudes, and the consequences for American society, were strikingly similar to the changes that had taken place during the prolonged agricultural depression of the 1880s and early 1890s, and again during the stop-and-go decade that followed World War I. Movement toward opening American society, either domestically or with respect to outsiders, mostly slowed or ceased . . . Attitudes among average citizens now forced to question the security of their own economic position and made even more anxious for their children's, became less generous and less tolerant . . . In each case large numbers of people have come to believe that some hidden, purposeful cabal must be at fault, and only its defeat can restore the America they love and of which they feel a part. And in each case as they sought that end, the openness and tolerance of our society, and our commitment to our democratic ideas, have suffered.[26]

As economic insecurity increased beginning in the mid-1970s, baby boomers led the national move away from inner-directed and toward outer-directed values. Those baby boomers who were born between 1945 and 1955, and who were between fifteen and twenty-five years old in 1970, came of age during a period of extraordinary prosperity. It is little surprise, then, that their postmaterialist values were so strongly oriented toward fulfillment. Likewise, it is little surprise that those Americans born between 1977 and 1989, who came of age during a period of both rising affluence and rising economic insecurity, today display so many more status-oriented values than their parents did.

Given these shifts, it also makes sense that Americans in the 1970s and 1980s became more insecure, economically and psychologically, and thus more *outer directed*. But we should be careful not to misunderstand this insecurity. Americans were not worried about their material survival — having enough to eat, having a roof over their heads — in the way that much of America was during the Great Depression. Rather, they were worried about their postmaterial survival. In comparison to the postwar era, Americans were insecure about their status and their place in what felt like an increasingly materialistic and dog-eat-dog world.

Rising status insecurity and the continuing move away from traditional forms of authority — from traditional family and church to governments and employers — all helped create the conditions for a powerful conservative backlash against abortion, same-sex marriage, gun control, and environmentalism. Insecure Americans seeking ways to meet their deeply felt needs for status direct their ire at immigrants, racial minorities, the poor, and others perceived as low status. And since the 1970s, conservatives have successfully harnessed resentments around everything from affirmative action to school busing to environmental laws. And while support for universal programs like Social Security and Medicare has remained relatively high, support for programs aimed at helping the poorest Americans has declined sharply.[27]

5.

By 1993, the economic fortunes for the average American started to improve. Per capita income grew 2.6 percent per year, and the median family income rose from $46,300 in 1993 to $54,200 in 2000. Whereas the median wage for a thirty-year-old man had declined 24 percent over the previous twenty years, his median wage rose 19 percent from 1995 to 2004. Unemployment reached a thirty-year low.

Given the powerful correlation between rising prosperity and the emergence of higher postmaterialist values, it would seem to follow that the increasing prosperity of the 1993 to 2003 period would be accompanied by a turn back toward fulfillment values. And indeed, this is what Benjamin Friedman believes happened. Friedman points to the quieting down of controversies over affirmative action, the liberalization of trade, military action in Yugoslavia and Somalia for humanitarian reasons, increased spending on education, and more liberal attitudes toward immigration.

But Friedman offers no evidence that any of the above is indicative of a shift in social values. The North American Free Trade Agreement (NAFTA) passed in Congress in 1993, when the economic recovery was just beginning, because President Clinton pushed it through, not because an increasingly prosperous and secure public had suddenly embraced trade liberalization. Notably, in 2005, after twelve years of solid economic growth (only briefly interrupted by a small recession), Congress rejected an expansion of NAFTA to the rest of Latin America.

Affirmative action was dramatically rolled back in a voter initiative in California in 1996 — a banner year for economic growth. Clinton's mend-it-don't-end-it position was politically deft, but it did not indicate (nor does Friedman suggest it did) more liberal attitudes toward racial equity. The only opinion research Friedman cites to substantiate his claim about more liberal attitudes toward the less fortunate is in regard to immigration. But the survey questions Friedman cites are entirely related to attitudes about the connection between immigrants and the economy — not about immigrants themselves. This difference is crucial. In an economy approaching full employment, as it was in the late 1990s, the vast majority of Americans no longer found the argument that immigrants were stealing American jobs particularly persuasive.

Our research suggests a more complicated picture: xenophobia has been rising, not declining, since 1992, perhaps the result of increasing *status* insecurity rather than increasing *job* insecurity. The percentage of people who agreed with the statement "Overall, there is too much immigration. It threatens the purity of our country" went from 45 percent in 1992 to 50 percent in 2004. The percentage of

Americans who agreed with the statement "Immigrants of different races and ethnic groups should set aside their cultural backgrounds and try to blend into the American culture" went from 50 percent in 1992 to 58 percent in 2004. And the percentage of Americans who agreed that "nonwhites should not be allowed to immigrate to our country" rose from 16 to 25 percent during the same period.[28]

In his 2005 book *American Backlash*, the social-values researcher (and our business partner) Michael Adams points out that the "Darwinist and exclusion" values of acceptance of violence, just deserts, ecological fatalism, sexism, sexual permissiveness, and xenophobia have all increased. Acceptance of violence is particularly noteworthy for the fact that it was the fastest-growing value in America from 1993 to 2004. This trend is seen in all classes among all races but is especially acute among the youngest Americans — an troubling indication of the evolution of social values.[29]

6.

It used to be that the wealthiest Americans were the fattest Americans. Today, the fattest Americans are the poorest Americans.[30] When reporters look for personal stories to illustrate obesity statistics they tend to travel to settings of concentrated poverty, like East Harlem or the Appalachian Mountains.[31]

Because obesity so disproportionately affects poor Americans, we tend to think of it as a problem of *poverty.* But if it is, then it is a very different kind of poverty than the poverty that afflicted the residents of Harlem and the Appalachian Mountains seventy years ago. Contrast the food situation of the poor during the Great Depression with the food situation in the Appalachian Mountains of Kentucky today.

Down the road live two brothers and a father who share a small house with an unfinished floor and a hole in the bathroom wall.

One of the brothers, James White, 42, weighs 340 pounds and has high blood pressure, but still eats mostly fatty foods, such as fried chicken and pork chops. His older brother Eddie, 46, had a six-way bypass last year . . . Their father, Fred "Justin" White, 70, has heart disease, chronic kidney failure and Parkinson's disease.[32]

Fried chicken and pork chops — these are foods that the poor in places like Nairobi, Port-au-Prince, and Rio de Janeiro can only dream about. In comparison with Americans during the Great Depression, the hundreds of millions of humans who still suffer from hunger, and 99 percent of all humans who have ever walked the earth, the poorest Americans today eat like kings. Today's poor are yesterday's middle class in terms of income and consumption.

And when they overeat, it is often because eating offers a feeling of comfort and security. Greg Critser, author of *Fat Land: How Americans Became the Fattest People in the World*, describes the situation of the poorest Americans in this way:

They are more likely to experience disruptions in health care, interruptions in income. Food, and the ability to buy it, comes in similar episodes — periods of feeling flush, periods of being on the brink of an empty pantry. The impulse is to eat for today, tomorrow being a tentative proposition at best.[33]

Critser is right that poorer Americans have more episodic lives, and that some overeat today fearing that they won't eat tomorrow. But it's no longer the case that the poorest Americans' ability to buy and eat large quantities of food, whenever and wherever they want, is a "tentative proposition at best." If it were, then the poorest Americans would not be the fattest Americans.

Understanding obesity involves understanding the consequences of prosperity. Food today is cheaper than it has ever been. As part of our transition from an industrial to a postindustrial economy, Americans are using and moving their bodies less than they ever have. Jobs require less physical labor and longer commutes. When not working or commuting, Americans are spending a record number of hours sitting in front of their televisions.

The obesity epidemic is not just the result of material *prosperity* but also the result of postmaterial *insecurity*. The United States, in contrast to other wealthy countries, has failed to create a new social contract appropriate for the postindustrial age. When Americans today fear for their economic lives, what they fear is not starvation or dying from lack of medical treatment but rather losing their status, their community, and their quality of life.

Materialist liberalism, as epitomized by the New Deal and the prosperity of the postwar period, was just what was needed to deal with hunger and material deprivation. But it provides no framework for understanding or addressing problems like obesity. In failing to create a politics that speaks to Americans' needs for security, status, and fulfillment, materialist liberalism is today foundering on the shoals of unmet postmaterialist anxieties and desires.

7.

When it comes to the state of the American economy, the left insists that Americans are more insecure while the right insists that they are more affluent. The truth is, Americans are both.

How is it possible for people to become richer and more insecure at the same time? Part of the answer is that American society has undergone a profound transformation, from an industrial to a postindustrial economy. The economic boom of the late 1990s was characterized by both extraordinary economic expansion *and* massive corporate downsizing and outsourcing. The new jobs, however high paying, were less secure. While a few individuals might hit the jackpot playing the Internet startup lottery, lifetime employment on the General Motors assembly line where your father spent his entire career

was a thing of the past. The right wing continued its thirty-year attack on the industrial-era social safety net begun in the Reagan years.
And while Clinton-era liberals recognized the need to replace it with
a new postindustrial safety net, they made very little headway actually
doing so.

Thanks to the conservative movement's dismantling of the social safety net, Yale professor Jacob Hacker notes, "economic *instability* of American families has actually risen much faster than economic *inequality*." Personal bankruptcies went from 290,000 in 1980
to two million in 2005. The mortgage foreclosure rate increased five-
fold. Eighty million Americans went without health insurance at
one point or another between 2002 and 2003. And what's been perhaps most nerve-racking for families is that income volatility since
the early 1990s has been two to five times higher than it was in the
early seventies.[34]

Still, Americans were not worried about their *material* survival,
as so many liberals tend to believe, but rather about their *postmaterial*
survival: fitting in, showing off, keeping up with the Joneses, and
getting their children the best education they could afford. When
asked what constituted the good life, far more people named material
goods in 1995 than did in 1971. Those who said "a happy marriage"
went from 84 to 77 percent. Those who said "a second car" went from
30 to 41 percent, "a swimming pool" from 14 to 29 percent, and "a va-
cation home" from 19 to 35 percent. And yet the percentage that
thought they had "a very good chance of achieving 'the good life'"
dropped from 35 to 23 percent.[35]

Poor Americans during the Great Depression developed a sense
of solidarity, in large measure because there was no denying their fall.
Materialist appeals worked because people could see that they weren't
alone. Fortunes evaporated, in some cases literally overnight. One-
third of the work force was unemployed. Standing in a breadline,
pleading with your neighbors for food, and walking all over town
looking for work — these were public acts of poverty. If waiting in a

breadline was humiliating, then you could at least be comforted by the fact that so many of your neighbors were standing in line with you. "When one-fourth of the entire country's labor force is unemployed at once, and a much larger fraction suffers joblessness at one time or another during the course of the crisis," Friedman noted, "people have a tendency to think that, whatever is happening, they are in it together."[36]

But thanks to cheap foreign labor, new supply-chain efficiencies, and increased productivity, even the poorest American saw his purchasing power continue to rise since the 1970s. The prices of many consumer products, from televisions to video games to air conditioners, fell dramatically, allowing the poorest Americans to purchase products that just a few decades earlier would have been prohibitively expensive.

As a result, when Americans look around today, they don't see the suffering of their fellow man, as they did during the Great Depression. They don't see the Jones family's credit card bills, mounting debt, and rising anxiety. Instead, they see the Jones family buying an even bigger SUV and moving into a larger house. An American today can be poised on the edge of bankruptcy and have a brand-new pickup truck in his driveway for all the neighbors to see.

Moreover, because of the intensification of media, marketing, advertising, and globalization, we have tended in recent decades to compare ourselves with far wealthier groups. "Today a person is more likely to be making comparisons with, or choose as a 'reference group,' people whose incomes are three, four, or five times his or her own," the economist Juliet Schor writes. "The result is that millions of us have become participants in a national culture of upscale spending."[37] Schor found that half of all Americans, including 27 percent of those making more than $100,000 a year, say they can't afford to buy everything they "need."[38] Almost all Americans say they want to be richer (85 percent of Americans say they aspire to be among the richest 20 percent) and only 15 percent aspire to be middle class.[39]

Today, many Americans find themselves caught in a vicious downward spiral masked by increased material affluence. More of us are working harder than ever, spending more hours commuting to and from work.[40] The average American household today spends about 13 percent of its after-tax income to service its debts — the highest point in twenty years. Most of it goes to house and car payments. But increasing amounts of it go to paying off credit card debts; the average household owes more than $8,000.[41] Six out of ten households that earn $50,000 to $100,000 a year are in credit card debt.[42]

Median household debt rose from $60,000 in 1990 to $100,000 in 2005, far outstripping the increase in median incomes.[43] For the first time since the Great Depression, the personal savings rate in 2005 went negative — which is to say, Americans spent more than they earned.[44] Meanwhile, total consumer debt reached $2.13 trillion in 2005, and home mortgage and home equity debt reached $8 trillion.[45]

On the one hand, increasing consumer and residential debt has allowed poorer Americans to enjoy lifestyles that were once considered upper middle class. It has allowed increased homeownership, which brings with it a greater investment in one's community, greater personal stability and security, and higher standards of living. By 2004, minority homeownership had gone above 50 percent for the first time in American history.[46] And of course, debt is a key part of the virtuous cycle of economic growth: debt increases consumption, which increases production, which increases employment, which increases consumption.

The problem is not high debt per se but rather high debt combined with low savings. Americans owed large debts after World War II, a period of rising prosperity, rising expectations, and rising security, but they also managed to put a good deal of money into savings, which has a hugely calming effect psychologically.

The debt Americans hold today is far more nerve-racking. Between 2004 and 2006 there was a 59 percent increase in one-year adjustable-rate mortgages.[47] The novelist Walter Kirn captures the high anxiety associated with this kind of debt.

> I find myself reading the papers nowadays with an especially anxious eye for looming short- to midterm problems that might somehow boost my payments to the moon. Unfortunately, like many in my position, I lack the sophistication to discern what sort of problems I need to be afraid of. A dirty bomb? Global warming? The bird flu? Will they help or harm me? Not as a human being, I mean, or as a citizen, but as a guy on a budget whose old frame house could, frankly, use a second bathroom, if he could only get the financing . . .
>
> Isn't the idea of a home to give a person a sanctuary from history? To grant him shelter from the worldwide storm? Maybe in theory. Maybe way back when. Today, however, our "residential properties" (because of the ways we have arranged to pay for them) are a source of our vulnerability.[48]

8.

The problems of insecure affluence are exacerbated by America's failure to create a new social contract appropriate for our postindustrial economy. For the last twenty-five years, conservatives have led the political effort to cut America's industrial-era social safety net. Democrats, progressives, and liberals have either uncritically resisted those efforts or uncritically implemented them. What they haven't done is acknowledge the ways in which the new needs are postindustrial not industrial, postmaterial not material, in order to create a compelling agenda for a new social contract on everything from health care to retirement security to child care.

We live and work in vastly different ways than did our parents and grandparents. We are unlikely to stay at one job for our entire careers. We are as likely to move across the country as across town, and we are likely to do so several times during our lives. We are more likely to work in the knowledge or service economies than the manufacturing economy and do so in workplaces that are less hierarchical and more flexible, as well as more intellectually demanding, than the assembly-line jobs of old. The postindustrial workplace offers much less fertile ground for unions than the workplaces of the industrial economy. And just as we more often identify up not down the class ladder, in today's knowledge, service, and creativity-centered workplace hierarchies, we are more likely to identify with management, not labor. Even if we don't like management, we aspire to it and have more opportunities to attain it than did our industrial-era ancestors.

We are also increasingly suspicious and distrustful of authority of all kinds. We demand increasingly unmediated relationships with power and authority. And we want greater flexibility from government that is consistent with the flexibility we demand in many other parts of our lives.

And we increasingly engage the world as consumers. By that we do not mean simply that Americans are increasingly materialistic, though we are certainly that. What we mean is that we engage more and more of our world through the mindset of the consumer. That mindset is more individualistic, transactional, and demanding, and we apply it to many more facets of our lives.

Liberals and progressives often see this demand for greater choice and individuality as a fall from a more beloved community, where we were more satisfied with what we had and were more caring toward our fellow man. Conservatives tell a similar story, but it is focused on when we were more responsible rather than more caring.

But from another perspective, the demand for greater choice is part of the long-term growth of prosperity and freedom in human civilizations. From this perspective, we haven't fallen from a golden age of community — we have risen from aristocratic dictatorships, patriarchal families, and hierarchical, authoritarian workplaces, all of which limited to just a few choices what we could do and who we could become.

Changing values and shifting identities have major implications for how Americans think about a new social contract. We use the term *social contract* rather than *social safety net* deliberately. The fact is, America still has a social safety net, despite conservatives' cuts to it. That safety net was constructed to meet the needs of Americans during the industrial era. Few Americans had retirement savings in 1930. Outside of unionized workplaces, few working-class Americans had health insurance in 1950. Public assistance programs were created at the height of the Great Depression, when roughly one-third of the work force was unemployed.

But today, the vast majority of Americans have health insurance and retirement savings. Close to 70 percent of Americans own their own homes, compared to 45 percent in 1950. More than half of all Americans own shares in the stock market, mostly through mutual funds and retirement accounts. And unemployment is considered to be a problem when it approaches 6 percent.

That nearly fifty million Americans lack health insurance and that Social Security may face a financial crisis when baby boomers retire — these are major problems to be sure, but ones of a different nature than the kinds Americans faced in the 1930s or the 1950s.

What prosperity and abundance bring is a demand for greater control, autonomy, flexibility, and choice. These are the needs that Americans now experience in their lives and to which a new social contract must speak. These are also the needs for postmaterial control and security that must be met in order to create the cultural conditions for creative thought, innovation, and inventiveness — the highest values of the postindustrial economy that in the coming years virtually every American will join. Not incidentally, they are also the conditions necessary to imagine and build the new clean-energy economy.

To understand the implications of all this, consider the case of health care. Most Americans today have reasonably adequate health insurance, but it is tied to their places of employment. The same is often true of their retirement. Anyone seeking to leave a job, start a new career, or move to a different part of the country risks leaving his or her health care and retirement savings, and the security they bring, behind. If it were easier for people to keep their benefits, they would have more choices in terms of what, where, and with whom they work, thus allowing more freedom and flexibility without raising serious concerns for their health and retirement security.

Moreover, America spends roughly 15 percent of its national wealth on health care, yet little of that expenditure is strongly correlated with better health outcomes. This is not simply the result of insurance and drug company profiteering. The medical establishment has refused to organize or discipline itself in ways that would make it accountable to actual health outcomes and best practices.

Meanwhile, our children are increasingly at risk of growing up to be morbidly obese. And medical consumers have little understanding of the efficacy of medical treatments or metrics by which to evaluate them and are increasingly fatalistic about their ability to control the factors that are the greatest drivers of long-term health outcomes, such as diet, exercise, stress, security, status, and social connectedness.

A high degree of control of one's life — whether in one's home, workplace, or community — is strongly correlated with both self-reported happiness and real health outcomes. Postindustrial Americans desire to be in control of their lives and creators of their futures, and this is what a new social contract for a new age must offer them.

That contract should not overlook the millions of Americans today who lack the most basic health and retirement security. These are preconditions to autonomy, control, and meaningful self-creation. Those Americans who "work hard and play by the rules," as President Clinton aptly put it, should be able to meet their basic material and security needs.

But we must not simply stop there, for to do so would be to ignore the real needs of the vast majority of Americans who work hard, play by the rules, and are more than able to meet their material needs. In a globalized economy, one's current job and current benefits should not be an obstacle to imagining and starting a new career and life. In a world of abundance, we must take more control over the food that we feed our children and ourselves. In a nation that spends 15 percent of its gross national product on health care, Americans need useful metrics to inform how we use medicine and take control of our health. In a country in which working Americans, through their pensions and retirement savings, now in large part own the means of production, the new fault line will increasingly be between management and ownership, not management and labor.

For many liberals, this rising demand for choice and possibility feels consumerist, selfish, and less communitarian than the class-based, redistributive populism of the New Deal and the Great Society. And in one sense that is true. We consume more than ever before, and the more we have the more we want. And in many ways we are lonelier, are more depressed, and have fewer close, personal relationships.

But at the same time, there is nothing wrong with our demand for increasing prosperity and freedom. Never before in human history have so many people cared about poverty, sickness, and the hardship of others, or for the nonhuman world, as they do today. Never before have women, gays, and minorities of all kinds had so much freedom. And never before have humans so widely considered one another — across national boundaries, skin colors, and language — to be members of the same species.

Americans today aspire as much to uncommon greatness as they do to the common good. They aspire to be unique, not common. None of this undermines empathy, compassion, or generosity. On the contrary, it is only when people are feeling in control, secure, and free to create their lives that they behave expansively and generously toward the collective.

The new social contract must thus provide a basis for people to seek individuation and self-creation. For liberals it means that we must recognize that Americans today are primarily motivated by aspirations for individual accomplishment and opportunity, and that a politics and agenda that speaks to those aspirations does not require that we abandon the communitarian context that made them possible. Indeed, it is necessary to recognize that our common needs have changed, so that we might imagine the new common preconditions required for Americans to realize their present-day aspirations. Seeing and seizing those opportunities means letting go of the older, materialist explanations of political behavior that emerged from the hardship of the Great Depression and the politics of the New Deal.

9.

There is perhaps no better case study of the political dangers in using an industrial-era mentality to create a postindustrial social contract than the repeated failed attempts by Democrats to fix health care.

Polls show that large majorities of Americans support guaranteeing all Americans access to affordable health care. Still, for forty years, health care advocates have been unable to win comprehensive, affordable health care for all Americans. Inaction on health care has been even more puzzling to Democrats than inaction on global warming is to environmentalists. That's because, whereas global warming ranks dead last in voters' priorities, health care usually ranks as a top concern, just after jobs and the economy, Iraq and terrorism. The problem is that as soon as health care advocates propose a specific policy solution, that consensus quickly falls apart.

Liberals tend to explain inaction on health care in the same way environmentalists explain inaction on global warming: corporate interests are corrupting the political process and deceiving voters. And to be sure, this is part of the problem. Insurance and drug companies give millions of dollars in campaign contributions every election cycle to curry favor with members of Congress. And they pull out all the stops to defeat serious reform efforts, such as President Clinton's health care reform proposal in 1994.

But in focus groups, we find the picture to be even more complex. When voters tell pollsters that they support affordable health care for everyone, what many are actually saying is that they think their health care is too expensive. And when we ask them to explain to us why that is, we find that they primarily experience the health care crisis as consumers — consumers who are being ripped off.[49] Their health care premiums and co-pays keep going up, they have a harder and harder time getting their insurers to approve the treatments they want, and they have less and less confidence that their medical coverage will actually be there when they need it. This is the health care crisis as most American voters experience it.

As morally wrong as it is that nearly 50 million Americans are uninsured, we should not forget that 250 million Americans *have* insurance — including the vast majority of voters. This is the place where health care reformers tend to be the most disconnected from the electorate. Reformers spend most of their time talking to the 250 million Americans who have health insurance about the 50 million who do not.

Aware of this disconnect, health care reformers attempt to blur the distinction by explaining to insured America how tenuous their health coverage might be. But further scaring insured Americans about the insecurity of their health coverage does not have the effect that health reform advocates intend. Instead of provoking altruism, generosity, and the demand for universal health coverage, scaring Americans about their health coverage provokes a zero-sum response. Insured, voting Americans become more likely to blame immigrants, the poor, the irresponsible, and the undeserving for skyrocketing health care costs. And they become more likely to take the attitude that if health care is too expensive already, increasing demand by adding more people into the system will make it even more costly, most especially to them.

Another approach would start by solving the health care crisis most salient to voters — namely the one that they experience every time they go to the doctor or pay their medical bills. Many health care advocates fear that if they solve the consumer crisis, the insured will not be motivated to support the health care needs of the uninsured. But research suggests the opposite: the more secure insured voters feel about their health care, the more generous they are likely to be in extending coverage to the uninsured. This means creating a politics where voters will feel assured that, for instance, their costs will not spiral out of control and their health care will be there when they need it, such as when they change jobs or get sick.

When it comes to proposing solutions, liberal reformers tend to be divided between those who want to push for universal coverage of children as a step toward universal coverage and those who want to make public health insurance programs such as Medicare available to everyone. But both approaches fail to understand how Americans reason about their health care.

While Americans today strongly support Medicare, they do so for *seniors,* not for themselves. In an America increasingly divided between the haves and the have-nots, public services are seen as being for the poor and those of low status: public transit for the poor, public assistance for the indigent, and public health care for the impoverished and elderly. Giving voters a choice between getting their health care from a government or a private health care provider is like giving Americans a choice between riding the bus and driving to work. Many Americans may still support making Medicare available to all, but they don't see it as a solution to their own health care needs.

Universal coverage for children also receives strong support from large majorities of Americans. And Congress may well pass legislation achieving that goal. But it may be that expanding health coverage to all children will actually set back efforts to cover all Americans. When voters in focus groups are asked whether, once children are covered, they would support extending coverage to the children's parents, many say they wouldn't.

Universal children's coverage is a relatively easy sell to Americans, because children are seen as *deserving*—a view that makes it more difficult, not easier, to extend coverage in the future to populations less easily characterized as deserving. In arguing for the expansion of health coverage to children, universal health care advocates, ironically, risk reinforcing the view that some people are more deserving of health coverage than others.

In keeping with what we know about how present-day Americans think about government and what they need from it, a successful strategy must first solve the health care crisis that Americans understand and experience every day. A reform strategy must make insured, voting America more, not less, secure about the status of their own health care. It must provide Americans with the kinds of choice and flexibility that they expect in all facets of their lives. It should position government as the enforcer of rules to ensure that everyone gets health care—not as the provider of services. And it must avoid reinforcing the widespread perception among the insured that those without health care are personally responsible for their plight and thus undeserving of assistance.

All of the above challenges industrial-era thinking about national health care and what must be done to achieve it. "Medicare for all" might have made good sense to Americans in the 1950s and 1960s when Medicare was first established, but it holds far less allure today. That a country as wealthy as the United States allows so many of its citizens to go uninsured is a travesty. And the costs and inefficiencies of the private health care system are a disaster. But we should not forget that, compared to the 1930s or the 1950s, the vast majority of Americans, more than 80 percent, *do* have health insurance. The result is that few Americans are interested in trading their private insurance for a public system or public insurance.

In focus groups, it is not unusual for Americans, in the midst of listing the many problems with the health care system, to pause and give thanks for their health. It is a moment when the participant reminds herself that she is strong, healthy, and lucky. It is also the mo-

ment when participants tend to be the most empathic and generous. Americans should be angry that we have a health care system
that allows insurers and drug companies to profiteer while millions
of Americans have no health security at all. But if, in the process of
getting angry, they do not also feel strong, healthy, and secure, their
anger will ultimately work against, not for, health care reform.

10.

In *What's the Matter with Kansas?* Tom Frank correctly identifies the
resentment of gays, intellectuals, and liberals as compensatory efforts
by the insecure to feel better about themselves. But telling working-
class Americans that they are fools is not the path to victory. About
the worst thing you can tell the economically insecure and the status
anxious is that they are victims.

   Kansas was received as a critique of moralizing but is itself the
ultimate morality tale. Frank fancies himself a populist but it's plain
that he can't stand the masses of people he grew up with. Frank writes
as though contempt flows only one way, from the backlashers to the
liberal elite, but the feeling is quite mutual. Frank wields pity like a
weapon, to club fools who forsake materialist rationality for postmaterialist morality.

   Whereas moral-values crusaders tell their followers that they
are spiritually rich and morally superior, materialist liberals tell their
followers that they are materially poor and intellectually inferior.
The liberal materialist's resentment narrative thus contains within
it its own contradiction: when Americans feel they are poor, victimized, and low status, their response is not to join a rainbow coalition
of other low-status victims. Rather, they seek to fulfill their needs
through materialistic status competitions or through putting others,
such as gays, Jews, and nonwhites, beneath them.

   Progressive interest groups trapped in the materialistic mindset
have long sought to make the Democratic Party the party of victims

and victimization at the hands of corporations and heartless Republicans. Environmental justice leaders insist to Americans of color that they are poor, low status, and contaminated, and then wonder why they are so irrelevant inside and outside of poor communities. Environmentalists blame human efforts to control nature for the ecological crisis — and then blame others, from corporate polluters to the media, for their political failures. And materialist liberals like Frank insist that Kansans are poor fools and then scratch their heads at the exodus of working-class voters from the Democratic Party.

11.

Frank characterizes right-wing nostalgia for a halcyon *Leave It to Beaver* past as little more than an irrational yearning for the protective womb of childhood. It is thus more than a little ironic that he fills his book with nostalgic visions of a progressive Kansas of the populist era of the 1890s and the New Deal era of the 1930s — times when, Frank believes, the people of Kansas rationally acted upon their material self-interests. Frank ends his book with a eulogy for the Kansas of FDR's New Deal and President Johnson's Great Society.

> It was a "workers' paradise," my dad remembers now; the ranch homes and split levels housed the families of appliance salesmen, auto mechanics, and junior engineers at the giant Bendix plant just across the state line: upbeat people, guys with GI Bill educations and color TVs in massive fake-mahogany cabinets.[50]

Frank describes how the New Deal markers described in Kansas City's 1939 WPA guide can no longer be found. A tall tower, the public library, the meatpacking plants — all are gone. Frank calls what exists today a "disfigured landscape."

"I heard striking tales of this tragically inverted form of class consciousness," Frank writes, "of the numerous people who lost their cable TV because of nonpayment but who nevertheless sported Bush stickers on their cars."[51] Such tales are striking indeed — but more for what they tell us about the fall of materialist liberalism than for what they tell us about postscarcity Kansas. There are tragedies in Kansas, to be sure, but they have nothing to do with losing one's cable subscription due to nonpayment. What was once a stirring vision of courage and vigor in the face of starvation has become a narrative of resentment built from lost cable subscriptions and fallen New Deal towers.

As surely as apocalyptic environmentalists believe that the betrayal of nature leads to global warming, Frank narrates Kansas's fall from New Deal liberalism in millenarian terms. *Kansas* is a fall narrative, complete with the state's expulsion from the garden of liberalism:

> As you cast your eyes back over this vanished Midwest, this landscape of lost brotherhood and forgotten pride, you can't help but wonder how much farther it's all going to go. How many of those old, warm associations are we willing to dissolve? How much more of the "garden of the world" will we abandon to sterility and decay?
>
> My guess is, quite a bit. The fever-dream of martyrdom that Kansas follows today has every bit as much power as John Brown's dream of justice and human fraternity. And even if the state must sacrifice it all — its cities and its industry, its farms and its small towns, all its thoughts and all its doings — the brilliance of the mirage will not fade. Kansas is ready to lead us singing into the apocalypse.[52]

"Singing into the apocalypse" is great millenarian prose. The problem is that there is no such thing as great millenarian politics. The stories that the left tells about America in decline and ecological collapse and that the right tells about religious wars and apocalypse are all narratives of resentment motivated by the desire to annihilate one's enemies — to see them *get what's coming to them.* They are all revenge fantasies.

12.

In America, the political left and political right have conspired to create a culture and politics of victimization, and all the benefits of resentment and cynicism have accrued to the right. That's because resentment and apocalypse are weapons that can be used only to advance a politics of resentment and apocalypse. They are the weapons of the reactionary and the conservative — of people who fear and resist the future. Just as environmentalists believe they can create a great ecological politics out of apocalypse, liberals believe they can create a great progressive politics out of resentment; they cannot. Grievance and victimization make us smaller and less generous and can thus serve only reactionaries and conservatives.

As liberals and environmentalists lost political power, they abandoned a politics of the strong, aspiring, and fulfilled for a politics of the weak, aggrieved, and resentful. The unique circumstances of the Great Depression — a dramatic, collective, and public fall from prosperity — are not being repeated today, nor are they likely to be repeated anytime soon. Today's reality of insecure affluence is a very different burden.

It is time for us to draw a new fault line through American political life, one that divides those dedicated to a politics of resentment, limits, and victimization from those dedicated to a politics of gratitude, possibility, and overcoming. The challenge for American liberals and environmentalists isn't to convince the American people that they are poor, insecure, and low status but rather the opposite: to speak to their wealth, security, and high status. It is this posture that motivates our higher aspirations for fulfillment. The way to get insecure Americans to embrace an expansive, generous, and progressive politics is not to tell them they are weak but rather to point out all the ways in which they are strong.

8

BELONGING

AND FULFILLMENT

MIDWAY THROUGH *What's the Matter with Kansas?*, Tom Frank visits Tim Golba, a leader of the state's grassroots, working-class conservative Republican movement known as the Cons. The Cons stand in opposition to Kansas's upper-crust Republican moderates, the Mods. The Cons are motivated largely by evangelical Christian concerns about abortion, homosexuality, prayer in school, materialistic status competitions, and sexual hedonism. They are also motivated by gun rights and tax cuts, which broadens their appeal to libertarian conservatives.

Frank portrays Golba as a monk who has taken a vow of poverty:

> Ignoring one's economic self-interest may seem like a suicidal move to you and me, but viewed a different way it is an act of noble self-denial; a sacrifice for a holier cause. Golba's monastic lifestyle reaffirms the impression: this is a man who has turned his back on the comforts of our civilization, who has transcended the material.[1]

But Golba is no deprived monk. He is a factory worker and simply not poor by American, much less by Brazilian or Great Depression, standards. Golba earns more than enough as a line worker in the local soda-bottling factory to live comfortably. To the extent that Golba has "transcended the material," it is because he has met his material needs and has moved on to realizing his postmaterial ones. And while Golba is awfully religious, he is hardly monastic. Rather, Golba is a part of a vibrant evangelical community that he loves and that loves him back. There is nothing suicidal about Golba's postmaterialism; Golba is a man who has long since satisfied his physiological needs and has dedicated his life to meeting his normal, healthy desires for status and power.

It must have been Golba's neighborhood that convinced Frank that Golba was an impoverished fool. Frank makes it clear that the upper-class suburb he grew up in was much nicer than Golba's, which is "the sort of neighborhood that hasn't aged too well." Frank finds the architecture tacky. "The houses all hew to the same general design, with only a few cheap ornamental features — fake balconies, plywood fleurs-de-lis — to dress up the box and distinguish one from another."

Liberals like Frank should be admiring, not ridiculing, Golba's modestly postmaterialist lifestyle. What Golba should be criticized for is not ignoring his economic self-interest but for trying to reestablish an outmoded, authoritarian, and patriarchal social order that limits rather than expands human freedom. We might disagree with his views on many things, but in terms of life's big existential questions — love, consumption, and community — Golba has a great deal of it figured out.

Believing that men like Tim Golba should identify with the poor and oppressed because they live in modest homes and make modest wages is as degrading as it is counterproductive. We should not see Golba as a victim of false consciousness or merely the banker's tool but as a hardworking, secure, and strategic right-wing moralizer. In short, we should see Golba as a *worthy opponent.*

1.

Between 2000 and 2004, Ohio lost roughly 250,000 jobs, mostly high-paying factory jobs.[2] For materialist liberals at the time, this fact alone meant that the state would almost certainly throw its votes to the Democratic presidential candidate, John Kerry. What liberals didn't notice was that Ohio's evangelical churches were thriving — and the ranks of conservative Republicans swelling. Fearful for their economic futures, their status, and their lifestyles, Ohioans responded neither with New Deal materialism nor with Clinton-era neo-liberalism but with moral-values conservatism.

Republicans mobilized quietly through Ohio's churches, out of sight of the Democrats and environmentalists who were funding the grassroots mobilization for John Kerry's presidential campaign. Democrats went into Election Day confident of a victory because they could not see the postmaterialist anomaly that was the successful evangelical mobilization of economic insecurity.[3] Increasing insecurity, loneliness, and hypermaterialism has boosted the growth of evangelical mega-churches in places like new exurban communities from just 10 in 1970 to 880 today. Though they represent just 1 percent of all congregations, they're rapidly growing, while older churches are shrinking.[4]

The political implications of all this are enormous. A few weeks before the 2004 election, liberal groups had to send volunteers or, more often, paid staff from places like New York and California into battleground states like Ohio. Conservatives, by contrast, were able to activate their existing network of evangelical churches to add voter outreach to their schedules of church and charitable activities in their own neighborhoods. For a few weeks prior to the election, conservative evangelical and fundamentalist churches set up phone banks, knocked on doors, and engaged in other get-out-the-vote activities as part of their community service.[5]

Evangelical churches aren't growing because they are preaching a political message. They are growing because they have made an art and a science out of addressing the need for community, companionship, and meaning among Americans who would otherwise be alienated. *Salon*'s young political journalist Michelle Goldberg observed in

Kingdom Coming, her 2006 polemic against Christian nationalism, that evangelical political power emerges from the search for belonging, fulfillment, and purpose.

> I met many people eager to engage in passionate discussion about the meaning of life, and about how we understand morality and reality. I saw the searching spirit that motivates many who find their way to the movement, their existential longings, and craving for a place in the world.[6]

2.

If you want to understand how a conservative Christian minority achieved political dominance in America, contrast environmentalism with evangelical Christianity. Both tell stories about humankind's fall, one from Eden and the other from Nature. Both tell revenge fantasies about a future apocalypse that serves as punishment for humankind's sins against either God or Nature. And both reward true believers with the warm glow of feeling morally superior to nonbelievers.

There are, obviously, many differences as well. Evangelicals openly ground their authority in religion, which they acknowledge is nonrational and faith based. Environmentalists ground their authority in science, which claims to be rational and free of faith.

Apocalyptic evangelicals preach of the end of the world and the Rapture, when Christians will go to heaven and everyone else will remain in a kind of hell on earth. Apocalyptic environmentalists preach of global warming hell on earth but offer no equally large redemptive vision.

And whereas evangelicals, apocalyptic or not, speak to a whole range of individual and collective *possibilities* for career, family, and purpose — an aspect of the American evangelical tradition that secular liberals have long misunderstood — environmentalists speak to the need to *limit* and even sacrifice individual and collective aspirations.

At the end of *Kingdom Coming*, Goldberg calls for "rebuilding a culture of rationalism," but her heart isn't in that old dream. Indicating, perhaps, that our generation is not doomed to the rationalistic appeals common to liberal baby boomers, Goldberg, who was born in 1975, writes, "a rational politics cannot promise the national restoration so many seem to long for."[7] Happily, she jettisons their materialism, too: "I don't think economic populism will do much to neutralize the religious right. Cultural interests are real interests, and many drives are stronger than material ones."[8]

3.

Almost two centuries ago, the French aristocrat Alexis de Tocqueville famously wrote about the curious American practices of freedom in a new kind of nation in a new world. Tocqueville admired the aspirational quality of the American character, but also worried about its dark side: greed, alienation, loneliness, and what would later be described by another foreign sociologist as a kind of one-dimensionality.[9]

What inspired Tocqueville about Americans — and what he saw as the antidote to our manic materialism — was our passion for self-transcendence through religious and community life. For Tocqueville, individual self-interest became the common good through churches and civic associations.

> Americans group together to hold fêtes, found seminaries, build inns, construct churches, distribute books, dispatch missionaries to the antipodes . . . If they wish to highlight a truth or develop an opinion by the encouragement of a great example, they form an association.[10]

Tocqueville's spirits were plainly buoyed by these ways of belonging, which he saw as incomparably superior both to the bureaucracies of the modern state and to the ancient aristocracies. He saw a kindness and solidarity in everyday interactions that were strikingly different from the formal systems of obligations in aristocratic France.

Virtually every significant social movement in American history has emerged directly from within religions, churches, and other community institutions or associations: abolitionists from Protestant churches; labor unions from factories and immigrant social networks; women's suffrage, temperance, and progressive reform movements from women's social groups; the civil rights movement from black churches; and antiwar, anti-apartheid, and anti-sweatshop movements from universities.

When one looks back on the progressive social movements of the last two centuries, what is striking is the profound meaning and transcendence they offered their participants, both as individuals and as part of a collective. They met postmaterial needs for esteem, status, and self-creation, while helping others achieve their more basic material and freedom needs, as in the case of the labor, civil rights, and women's movements. They sang and marched. They formed lifelong friendships and marriages. And they were deeply embedded within the communities in which most of the participants were born and would spend their entire lives.

But as those social movements succeeded and matured, and as American society transformed itself from industrial to postindustrial, from urban to suburban, and from material to postmaterial, many of the civic institutions that gave birth to those movements began to wither away. The decline of the industrial economy brought with it a concomitant decline in union membership. Social and economic mobility scattered ethnic diasporas throughout the country, thereby undermining ethnic identity and the old ethnic social networks as a suitable basis for organizing politically. Geographic mobility and suburbanization undermined community organizations. Increasing freedom and mobility for girls and women reduced the appeal of organizations like the Junior League. And while membership in evangelical churches has significantly increased, overall church attendance has declined, most especially among the old mainstream denominations that were the basis of many of the traditional liberal and progressive social movements of the past.

At the same time, the grand and transformative goals that inspired those movements were largely accomplished. Workers won the right to organize, and many of the workplace rights they had fought for — the minimum wage, the forty-hour week, occupational safety — were enshrined as the laws of the land. Women and minority groups were enfranchised and legal; institutionalized discrimination on the basis of race and sex was ended. The old stories — of the days of lynchings and Jim Crow, of kept women and back-alley abortions, and of child factory workers asleep on their feet and unable to stop the production lines — lost their ability to inspire or transform and their power as metaphors for the new challenges faced by a modern and affluent nation.

As those early- and midcentury social movements became mature, successful, and exhausted ideologically, as the social institutions from which they were born foundered, and as their agenda shifted from the broadly transformational to the narrowly issue based, they became interest groups. As such, they operate in the same way as other private interest groups, from the Chamber of Commerce to the National Association of Broadcasters. As they evolved into interest groups and their participants became once-a-year dues-paying members, these institutions lost their power to facilitate self-creation, transcendence, and the meeting of other postmaterial needs.

4.

The decline of the civic affiliations that undergirded the progressive social movements of the last century was no anomaly and was not limited to those social movements. America's social and civic disengagement has been well documented for several decades, most notably in recent years by Harvard professor of public policy Robert Putnam, whose *Bowling Alone,* published in 2000, became required reading for foundation presidents, columnists, and members of Congress alike.

Putnam argued that four factors were driving the decline: longer work hours and the entrance of women into the work force; suburban sprawl, which puts greater distance between family, friends, and work; increased television watching and media consumption; and the rise of the me-oriented generations and the decline of the more civic-minded World War II generation.[11]

Putnam documented the loss of what sociologists call "strong ties": deep, long-lasting connections that help people define themselves within a supportive community. The decline of these strong ties appears to be continuing apace: a 2006 study found that the number of friends or family the average American had to confide their most personal hopes and concerns in declined by nearly one-third. It also found that the number of Americans who say they have no one with whom to discuss important matters more than doubled.[12]

Affluence, social mobility, and the rise of the mass media and consumer cultures in the postwar era created unprecedented new opportunities for Americans to imagine and pursue new futures, possibilities, and identities for themselves. But these new opportunities had a downside: rising social isolation and alienation as Americans increasingly left behind family, community, and traditional social institutions in pursuit of their dreams.

With the decline of the old industrial economy, increasing numbers of Americans, whether by choice or necessity, left the tight-knit communities of the industrial era. For the better part of a century, most Americans had lived in communities that were urban and ethnic (although, notably, not multicultural), worshipped at mainstream denominations entwined in community traditions, and worked on the production lines of manufacturing behemoths that tended to dominate the local economy. Americans were more likely to stay in the communities where they had been born and raised. They were more likely to be surrounded by their extended family and community. They often worked an entire career at the same job or manufacturing plant. The person you worked next to on the assembly line also likely lived in your neighborhood, worshipped at your church, and sat on the stool next to you when you drank at the local union hall. The industrial economy provided a stability that helped create and maintain strong ties.

Affluence, modernization, and the demands of the new service and knowledge economies also brought a concomitant decline in orientations toward traditional forms of authority and social norms and institutions. Whether in the workplace, houses of worship, or politics, Americans increasingly demanded more unmediated relationships with power and authority and became less willing to accept traditional social norms inherited from their parents and grandparents.

In the end, the social capital and strong ties of the industrial economy were as doomed as the copper smelters, steel plants, and assembly lines that gave birth to them. It is no accident that the cities in which Putnam found the greatest social capital still intact were dying industrial towns, such as Milwaukee and Pittsburgh, and the cities where Putnam found it most absent were in the new economy towns, such as Seattle, San Francisco, New York, and Boston. Whatever the future may hold for social organization and affiliation in America, one thing is almost certain: Americans are not going back to the bowling leagues of Milwaukee and Pittsburgh.

5.

In documenting the decline of social capital, Putnam directed our attention to something important, but he also missed something equally profound, and that was the rise of new kinds of social capital, networks, and social ties. In *The Rise of the Creative Class,* Richard

Florida documented the increasing numbers of Americans working in the new knowledge and service economies who "share a common creative ethos that values creativity, individuality, difference, and merit."[13] These Americans, Florida found, have fewer of the strong social ties that Putnam valorizes. But they also have many more "weak ties," which are precisely what the creative class wants and needs to break free of more traditional, stifling forms of community.

> Traditional notions of what it meant to be a close, cohesive community and society tend to inhibit economic growth and innovation. Where strong ties among people were once important, weak ties are now more effective. Where old social structures were once nurturing, now they are restricting . . . People want diversity, low entry barriers and the ability to be themselves.[14]

The seminal articulation of this phenomenon was in sociologist Mark Granovetter's 1973 paper "The Strength of Weak Ties." In it, Granovetter concluded that weak ties help far more than strong ties in getting a job, innovating, or starting a new business. Where strong ties offer depth, weak ties offer breadth.[15] There is even evidence that the revered strong ties of the older industrial economy often served to exclude outsiders, whether immigrants or Americans migrating from outside the area, from employment and other economic opportunities. "Weak ties," Florida writes, "allow us to mobilize more resources and more possibilities for ourselves and others, and expose us to novel ideas that are the source of creativity."[16]

Florida draws a contrast between what has become known as blue America and red America, the former a "cosmopolitan admixture of high-tech people, bohemians, scientists and engineers, the media and the professions" and the latter a "more close-knit, church-based, older civic society of working people and rural dwellers."[17] And while he laments the excessive alienation, materialism, and individualism in places like Silicon Valley, he also finds hope in new creative centers, such as Seattle, Toronto, and Dublin, which have both the vibrancy of the new economy and the charm and character of old cities.

In his advice on how cities can become creative centers, Florida, a professor of public policy and urban development, argues for the three Ts: technology, talent, and tolerance. For cities to attract talent they must have a high quality of life, one characterized not by *Leave It to Beaver*–type suburban conformity but rather by density and diversity. The creative class overwhelmingly wants to live in cities where families can intermingle with bohemians, where immigrant-owned shops, restaurants, and coffeehouses provide a new kind of public space, and where nobody bats an eye at the gay couple holding hands near the playground.[18]

And while none of this is to suggest that strong ties are bad and weak ties are good, or that the new economy and the creative class are the cure for all of America's ills, it does suggest that the future portends more weak ties and fewer strong ones, the continuing decline of traditional forms of social capital, and an increasing need to create new forms of social capital that are appropriate for our age.

6.

Few liberals or environmentalists had heard of the Reverend Rick Warren before he signed a letter in early 2006, along with eighty-five other evangelical ministers, supporting action on global warming. Indeed, anyone who doubts how wide the gulf between red and blue America has become should consider that Warren had sold *23 million copies* of *The Purpose-Driven Life* before even the most well-read of liberals learned that he had become the Billy Graham of the early twenty-first century.

Warren's success is largely the result of having created a church that helps people get their needs for belonging and fulfillment met. It's located in a part of Orange County, California, that Putnam calls "a virtual desert in social-capital terms" — but what Florida would characterize as a community grounded more in weak ties than strong ones. Saddleback is a postmaterialist church for a postindustrial economy and, as such, it holds many lessons for how to create pre-political institutions that address our most strongly felt needs.

Before starting Saddleback, Warren went door-to-door in his neighborhood, talking to potential congregants. Many told him that they didn't go to church because they found the music square, the preaching irrelevant, and the children's programs lacking. So Warren started preaching, often in a Hawaiian shirt, about what really matters to people. His speeches were funny and free of fire and brimstone. He preached about things like how to deal with emotionally difficult people, how to be a good spouse, and how to simplify one's life. And Warren made it easy for people to attend Sunday services, offering convenient parking and an engaging Sunday school. In an article for *The New Yorker*, Malcolm Gladwell, the author of *The Tipping Point*, noted that "the doors are open, so anyone can slip in or out, at any time, in the anonymity of the enormous crowds."[19]

About one-third of Saddleback members belong to a small group of other congregants who all have something in common, such as working at Cisco Systems, being deaf, having a loved one in prison, or battling breast cancer.[20] These small groups become the Saddlebacker's extended family. Gladwell quotes Putnam:

> There were these eight people and they were all mountain bikers — mountain bikers for God. They go biking together and they are one another's best friends. If one person's wife gets breast cancer, he can go to the others for support. If someone loses a job, the others are there for him. They are deeply best friends in a larger social context where it is hard to find a best friend.[21]

These small groups are the church more than the building where services are held. In signing up for a small group one must commit to participate for roughly two months, but church officials say that

70 percent of the members continue for years beyond that. Today, an estimated *forty million Americans* are involved in a religious small group. What Warren has done, and done brilliantly, is maximize the strength of weak ties through his congregation, the ultimate hub of multiple networks, and convert some of those weak ties into strong ties through his small groups. "People are not looking for friendly churches," Warren says on his Web site, Pastors.com, "they are looking for friends."[22]

Traditional evangelical preachers disapprove. They accuse Warren of selling the Word of God like a consumer product. They point accusing fingers at the large flat-screen TVs that allow congregants in the back rows to see the service, which is accompanied by light rock music.[23] But this is just a cover for what they really disapprove of: Warren's success in creating a genuinely beloved community that allows for far more individuality and personal freedom than did the churches built by Billy Graham, Jerry Falwell, or Pat Robertson.

Warren has long spoken out against the spiritual abasement of materialism, and his expansive agenda is aimed at helping people deal with both postmaterial alienation and material deprivation. In thinking about what to do with his newfound wealth, Warren turned to Psalm 72, where the wise King Solomon, the richest man in the world, asks God for more influence and power — so that he could help people to overcome their poverty and suffering. Shortly after reading that, Warren says that he and his wife decided to give away 90 percent of the tens of millions of dollars they had earned from *The Purpose-Driven Life*. Inspired, perhaps, by Bono, Bill Gates, and Warren Buffet, who have dedicated their lives and much of their fortunes to compassionate action, the Warrens have funded microcredit programs and AIDS treatment in Africa.

Awed by the American way of belonging, Tocqueville asked, "What political power could ever substitute for the countless small enterprises which American citizens carry out daily with the help of associations?"[24] Gladwell quotes Warren making a similar point — but this time in a global context.

I can take you to thousands of villages where they don't have a school. They don't have a grocery store, don't have a fire department. But they have a church. They have a pastor. They have volunteers. The problem today is distribution. In the tsunami, millions of dollars of foodstuffs piled up on the shores and people couldn't get it into the places that needed it because they didn't have a network. Well, the biggest distribution network in the world is local churches . . . Put together, they could be a force for good.[25]

7.

Throughout this book we have criticized the ways in which environmentalists treat nature and science as a *religion,* which we believe lies behind environmentalism's ideological orthodoxies — its pollution paradigm, its politics of limits, and its policy literalism — and which prevents environmentalists from achieving their goals. But here we consider the ways in which environmentalism doesn't work enough like a *church.*

The average environmentalist — or environmentally minded person — enacts her environmentalist identity through particular practices, whether talking to friends and occasionally proselytizing, using new light bulbs, buying green products such as organic foods and hybrid cars, and giving money to groups such as the Sierra Club, the National Resources Defense Council (NRDC), or Greenpeace.

At first blush, this is similar to the ways evangelicals enact their Christianity. Evangelicals talk to friends and occasionally proselytize. They buy products in line with their values and occasionally boycott companies for being anti-family, pro-gay, or insufficiently Christian, such as in the case of Wal-Mart, which, for a brief period of time, wished its customers the blasphemously secular "Happy Holidays" instead of "Merry Christmas." And they give money to evangelical causes.

But when it comes to both the breadth and depth of membership and fundraising, the differences between evangelicals and environmentalists shine through. The largest environmental advocacy groups lobbying on global warming, such as the Sierra Club and the NRDC, have well under one million members each. Their political power, such as it is, comes from speaking for the 60 to 70 percent of the public they claim are strongly supportive of their legislative agenda.

One of the problems, as we have seen throughout this book, is that while public support for the environmental agenda is broad, it is also frightfully shallow. This has implications not just for the challenge of passing legislation but also for the inseparable challenge of building a movement.

The environmental movement quickly shrinks when one defines its members as those who give financial, not just verbal, support for the cause. Seventy percent of Americans will tell pollsters when asked that they support environmental objectives such as clean air and clean water, but only somewhere between ten and twenty million Americans actively support environmental groups by giving money or taking actions such as writing letters or calling legislators. Even this group of active supporters turns out to be surprisingly modestly engaged. Most who give money give less than a hundred dollars. And writing a small check once or twice a year tends to be about all that they do. They do not attend meetings or protests. They do not have structured or consistent time when they engage with other environmentalists as environmentalists. And they do not have meaningful public expressions of their environmental identity other than occasionally showing off their hybrid cars.[26]

By contrast, the evangelical movement is large, growing, and phenomenally wealthy. *Serious* evangelicals — to be distinguished from those evangelicals who go to church only occasionally and give small sums of money when they do — contribute a large percentage of their income, often as a tithe (10 percent), to their local church. They also give money to national advocacy organizations, such as Focus on the Family or the Family Research Council. If the median household income for people in their prime working years (twenty-five to fifty-nine years old) is roughly $60,000, a serious evangelical from a median-income family gives $6,000 of that to his local church — about sixty times more than an environmentalist who writes a hundred-dollar check once a year.[27]

And that's the least of it. Arguably the most important difference between environmentalists and evangelicals is in the experiences the two groups create for themselves. The serious evangelical not only gives away money but also goes to church and participates in charitable activities. While evangelical Christianity is an individual *and* community experience, environmentalism is mostly an individual one. Outside of giving money and buying green products, few among even the most serious environmentalists ever actually do anything to manifest their environmentalist identities or to recruit others to join them. In short, while the evangelical identity is *thick,* the environmentalist identity is *thin.*

To the extent that environmentalists have meetings at all, they are more depressing than inspiring, focused more on stopping development than creating a beloved community. Drive across town to the local mega-church service and you'll likely find an energetic and vibrant righteousness that doesn't get to the dull work of door knocking and phone banking until well after the faithful have sung songs and felt the warming love of Jesus in their hearts.

8.

In order to create a politics that people want to be a part of, we need to take a step back and understand at a more fundamental level what makes people happy and fulfilled. The last two decades has seen an explosion of social science research into the new sciences of happiness and an attendant series of very fine books summarizing this research.[28]

One of the most surprising and relevant of these findings is that we are often at our happiest when we are in *flow* — that is, in the state of losing oneself in activities that require a level of concentration that is intense but not so intense as to become frustration. Studies have been conducted in which individuals carry small alarm clocks that remind them at various points in their day to *rate* their relative state of happiness. What the studies find is that people are often at their happiest when they are experiencing flow.

Another profoundly fulfilling activity is *service to others*, whether through building houses with Habitat for Humanity or helping somebody cross the street. Notably, these acts of compassion are also always acts of strength, and the positive feelings they engender are inextricably linked to our sense of feeling powerful while doing them.

Perhaps the most obvious finding, and the one most connected to belonging, is that happiness comes from *being with others*. Strong social ties are crucial not only to living a fulfilling life but also to living a healthy one. In fact, there is a large and growing body of research showing that strong social ties may offer greater health-protective effects than even quitting smoking, drinking less, or eating a healthy diet. And the lack of attachments appears to have the opposite effect, making us lonelier and more at risk of drug and alcohol abuse, suicide, depression, stress, and overall unhappiness.[29]

These social attachments are bound up with living a *meaningful life*. Nobody creates his life's purpose in a vacuum. The stories we tell about our lives, and the purposes we attach to our behaviors, are crucial to creating happiness.

In short, people go to places of worship to be with friends, experience flow through singing, feel powerful through service work, and create meaning for their lives. These activities help people meet their lower- and higher-order postmaterial needs for belonging, esteem, status, and self-creation. Given all of this, it's easy to see why millions of Americans give thousands of dollars to religious institutions, while a mere few hundred thousand Americans give only around a hundred dollars a year to environmental organizations.

9.

Progressives and environmentalists have become so accustomed to thinking of politics as little more than the three Ps — policies, politicians, and protest — that our discussion of belonging and fulfillment will no doubt strike some readers as digressive. Others will react by thinking that the comparison we have drawn between environmentalism and evangelicalism is unfair — after all, environmentalism is not a church and we should not expect it to behave as one.

But it is hard to imagine creating a politics powerful enough to transform the global energy economy that is not fundamentally grounded in people's lives. A bright new agenda backed by good science, effective marketing, and savvy lobbying will simply not be enough. Given all of this, creating a web of new pre-political associations may be the most important thing progressives, environmentalists, and postenvironmentalists can do to overcome ecological crises and social insecurity. "In democratic countries," Tocqueville noted, "the knowledge of how to form associations is the mother of all knowledge since the success of all the others depends upon it."[30]

Of course, if American evangelicals were as concerned about global warming and AIDS in Africa as they are about abortion, same-sex marriage, and prayer in school, there would be little need for a new politics and a set of institutions to go with it. But, unfortunately, they are not, and so the need remains. Despite the occasional rounds of media excitement over a few evangelical ministers endorsing greenhouse gas emission limits, every politician knows that as long as he takes the right position on things like abortion and same-sex marriage — the issues that matter most to evangelicals — it doesn't matter how reactionary his position is on global warming.

Indeed, when one of us challenged a leading evangelical environmentalist — Richard Cizik, the vice president of governmental affairs for the National Association of Evangelicals — on this point during a radio show, he admitted as much. "I happen to be among those who disagree with the Bush administration on climate change,"

Cizik said. "But when I went into the voting booth [in November 2004], yes, I had to make a decision in the competition of principles. And the sanctity of human life trumped stewardship of the environment in casting my vote for the presidency."[31] Which is to say, Cizik voted for Bush.

Both environmentalists and liberals see evangelical churches as *tools* to advance their policies when they should see them as *models* for how to construct a politics through communities of purpose, meaning, and vision. What we have been pointing to is not a recipe for creating a movement but rather a set of trends that could become the raw material of a new politics: the continuing importance of association and affiliation in American life; the rising importance and strength of weak ties; greater flexibility in work, family, political, and community life; and the rising importance of creativity to the knowledge economy.

What we are arguing against is the thin notion of politics which has taken hold of liberals and environmentalists: the notion that social and economic transformation can occur through better policies and marketing alone. Many environmental leaders today believe that they can advance a resonant politics on global warming if they use words like *stewardship* and quote relevant passages of Genesis.

Their faith is eerily matched by the faith among liberal leaders that they can advance their materialist agenda by declaring poverty a moral issue and talking endlessly about Jesus' Sermon on the Mount. Just as environmentalists believe they can galvanize the public by showing visual images connecting pollution to global warming, and liberals believe they can mobilize Kansans by demonstrating to them where their material self-interest lies, environmental, liberal, and Democratic leaders believe they can appeal to evangelicals by pointing to the passages in Scripture that are consistent with their political positions. But as long as they ignore the social and psychological needs that Christian evangelicalism helps believers to meet, no magical language of values or quotations of Scripture will have the political impact that environmental and liberal leaders desire.

10.

Given the way that our society has evolved, the values of life, liberty, and the pursuit of happiness have far different meanings for us than they did for Thomas Jefferson or Frederick Douglass. Happiness has evolved from meaning the absence of suffering and presence of pleasure to meaning a whole set of ideas about personal fulfillment, purpose, and connectedness to others. And freedom has evolved from meaning a set of rights to property and citizenship to meaning greater amounts of autonomy and choice and the opportunity to pursue one's happiness.

The problem is that in the 1970s and 1980s when preachers like Pat Robertson and Jerry Falwell were *politicizing* evangelical ways of belonging and fulfillment, the left was *privatizing* them. Falwell in particular made a concerted effort to unite evangelicals and fundamentalists and bring them out of their self-imposed political exile. And Republicans started seeing evangelical and fundamentalist churches as pre-political institutions that could be directed to flow into a conservative moral-values politics.[32]

Liberals and environmentalists have thus tended to be issue based and complaint based, while conservatives have tended to be values and needs based. Whereas liberals and environmentalists view politics as a means of dealing with issues — solving problems through legislation and the courts — evangelical conservatives view politics as yet one more domain for strengthening Judeo-Christian values in the culture. Liberals and environmentalists thus lack an overarching vision and a coherent ideological framework that could hold their agenda together and connect it to pre-political ways of living and institutions from which a movement might be built.

Recruiting the unsaved to go to church, protesting abortion clinics, working in soup kitchens, putting the Ten Commandments in courthouses, organizing Christian rock concerts, protesting (after confabulating) the "war on Christmas," mobilizing to continue life support for those whose brains have died, supporting Israel, creating evangelical X Games, banning the freedom of gays and lesbians to marry — all of these are various means to the same end of building a Christian nation. Evangelical politics, with its holistic vision for remaking society and human behavior, has been free of the policy literalism that has crippled environmentalism and liberalism.

It's not just the Christian right that turns private concerns into political demands. The antienvironmentalist right has successfully mobilized private resentments against environmentalist restrictions on off-road vehicles, from four-wheelers to snowmobiles, into a potent political force. And the National Rifle Association has successfully organized those Americans who enjoy hunting, target practice, and collecting firearms into a potent lobby whose most energetic members block even modest efforts to reduce death and injury from handguns.

The left complains that the religious right is seeking to make public beliefs and practices that are essentially private. But all politics is about determining what should be public, what should be private, and what should be banned altogether. For example, public schools today teach beliefs, such as evolution and human equality, that were once held privately by a very small, secular minority of scientists. They teach that all people are created equal and that humans evolved from apes. These secular beliefs were once no less *private* than the belief that some people will go to heaven and others to hell, or that God created humans out of whole cloth. The difference between these competing secular and religious beliefs is not that the former are views of the majority and the latter views of the minority, for only a minority of Americans today believe we evolved from apes, while a majority believe in heaven and hell. The difference is that the secular beliefs have, so far, won the political battle over what will and what won't be taught in public schools.

All beliefs are personal and pre-political before they become public and political. The civil rights and women's movements began with people seeing themselves and others around them differently than they had before. The social movements grew out of these new values and ways of being in the world. Politics is always a contest between what shall become public and what shall remain private.

Liberals believe that some things should remain personal and private (for example, the belief that Christians are God's elect) and others (the theory of evolution) should not. Things like the concept of human equality, the appreciation of other human cultures, and the theory of evolution by natural selection — these once-private beliefs have today become part of public education, and in the liberals' worldview, that is a good thing. The question is *which* private concerns should remain private and which ones should become political and public, and that is always a matter of perspective and often contested.

11.

The late Richard Rorty, perhaps the greatest American philosopher of the generation that came of age in the sixties, provided the most intellectually sophisticated justification for drawing a bright line between the personal and the political. He did so during the same time that Falwell and others began seeing evangelical and fundamentalist churches as pre-political institutions.

Rorty's argument was that there can be no synthesis of the two equally legitimate quests for self-creation and for social justice.

> The closest we will come to joining these two quests is to see the aim of a just and free society as letting its citizens be as privatistic, "irrationalist," and aestheticist as they please so long as they do it on their own time — causing no harm to others and using no resources needed by those less advantaged . . . The vocabulary of self-creation is necessarily private, unshared, unsuited to argument. The vocabulary of justice is necessarily public and shared, a medium for argumentative exchange.[33]

Rorty further argued that there can be no value more important to politics than enacting solidarity to prevent cruelty.

But social justice, solidarity, and cruelty prevention is depressingly thin porridge compared to the moral grandeur of evangelical Christianity — and unnecessarily so. One need not promote one's own *particular* search for personal excellence and private fulfillment to believe that certain institutions of belonging and practices of fulfillment are necessary elements of a powerful political movement.

And Rorty put too much emphasis on the negative politics of avoiding cruelty. It was almost as though he searched and searched for something at bottom to ground his politics; a fairly ironic move, given his own insistence on the contingency of moral and philosophical foundations. Almost everyone on the left today believes that war is sometimes necessary. But if wars are sometimes just they are also *always* cruel, and it is dangerous to imagine otherwise. Avoiding cruelty is thus less *useful* a standard from which to craft a politics than advancing prosperity, freedom, and the capacity for all humans to realize the potential for self-creation. "Achieving our country" — the name of a book he published in 1997 — is a far better basis for creating a new politics than avoiding cruelty, not just because the former is positive and the latter negative, but also because it speaks to possibility rather than to limits.

Having decided that private pursuits of happiness must not in any way be confused with politics, Rorty reinforced one of the worst of the left's contemporary errors of reason: that we must choose between individualism and social justice.

Nobody has yet suggested a viable leftist alternative to the civil religion of which Whitman and Dewey were prophets. That civil religion centered around taking advantage of traditional pride in American citizenship by substituting social justice for individual freedom as our country's principle goal. We were supposed to love our country because it showed promise of being kinder and more generous than other countries.[34]

But at various moments in American life, being a liberal has *also* meant loving our country because of the astonishing amounts of freedom it offers to us as individuals. This was the liberalism of Emerson in "Self-Reliance," Thoreau in *Walden,* and many a liberal boomer in the 1960s and 1970s. "The only sin," Rorty approvingly quoted Emerson as saying, "*is limitation*,"[35] and yet it was Rorty who repeatedly closes the door on the notion that our personal quests for authenticity can ever be captured by a truly *new* progressive politics, one that is far more expansive than simply building solidarity to defeat cruelty.

12.

Liberals today tend to talk of individualism and community as if they were opposites, often sounding like conservatives when they claim that the problems of postmaterialist alienation are the result of the same problems environmentalists blame for the ecological crisis: too many possibilities and too few limits.

While liberals are correct to be concerned with the decline of social intimacy, many have mistakenly conflated that decline with the proliferation of fundamental life choices. This misreading of the problem is ironic, because liberals and environmentalists are perhaps the most individualistic beings that have ever roamed the earth. They are seeking self-creation in their love, family, work, and community lives. They do not want to live under the moral strictures their grandparents lived under but want to decide for themselves how to live. All of this is quite healthy and normal and in no way inconsistent with being part of a community — even though particular ways of becoming an individual will be incompatible with particular communities.

Individualism emerged from the tribe. As human societies evolved, and people lived longer and healthier lives with greater wealth, their choices of whom to love, what to do, and what to believe increased dramatically. With this growing autonomy, people experienced greater potential to realize their unique selves. After we become different people through a process of self-creation, we find others who have been through a similar journey and we create new families, businesses, and communities with them. Becoming a creative and fulfilled individual in our postmaterialist society still requires belonging, in one way or another, to a community.

One can be part of a supportive community that helps one become authentic or one can be part of a community that constrains one's aspirations. One can as easily be alienated as authentic in one's community. Almost all of us have craved flight into solitude rather than remain part of a community we didn't choose and perhaps even despised. Better to quit your community and throw your social capital in the river than spend another minute pretending to be somebody you are not or no longer wish to be.

Whether you call it becoming self-reliant,[36] becoming who you are,[37] self-creating,[38] becoming a person,[39] self-expressing,[40] taking hold of one's life,[41] or becoming authentic,[42] this process could be a powerful pre-political basis for creating a new postmaterial politics. Seeking individual authenticity is not some quaint dynamic separate from the real work of social change — it *is* social change, and always has been. Women who decide to go to work despite their fathers and husbands; couples who decide to have sex before getting married; gays and lesbians who come out of the closet; and Christians who decide to be different kinds of Christians — all of these can be understood as forms of self-becoming, and all can become political to the extent that these individual acts, identities, and commitments determine our collective governance.

After achieving some degree of personal authenticity, postmaterialists are then capable of seeking out new communities governed by self-imposed rules. They are limits in the service of creating new possibilities. This regrounding is neither a rebellion against nor a blind acceptance of tradition, but rather the free taking up and remaking of traditions that are deemed useful in becoming the people we wish to become.[43]

In the past, communities were rooted in hierarchical forms of authority that were obeyed simply because they were meant to be obeyed. But the institutions serving postmaterialists, from governments to employers to evangelical churches like Warren's Saddleback, which leaves its doors open so congregants can arrive late or leave early, are all being forced to become more flexible and accommodating of the move toward greater freedom, autonomy, and possibility.

The problems liberals often point to as stemming from the decline of community are not the result of too many possibilities but of not enough. Our consumer culture has become expert at offering Americans an overwhelming array of hedonistic, materialistic, and pleasurable but short-lived choices, but it offers little in the way of possibilities for flow, service to others, self-mastery, belonging, or fulfillment — in short, the kinds of activities that provide happiness, accomplishment, and esteem. The problem is thus not only that liberals have failed to advance — and fight aggressively for — a vision of the common good, principally a new social contract appropriate for our postindustrial economy, but that they have also failed to speak to *the pursuit of uncommon greatness,* which is a fundamental aspect of the American character. Americans are motivated as individuals to seek their own unique purpose in life. To suggest that this drive is somehow anathema to community or social capital is to miss something fundamental about what it means to be — and become — an American.

13.

Liberals who simplistically juxtapose individual and communitarian aspirations fail to recognize the ways in which self-creation requires both individual aspiration and community. To be sure, much of self-creation is individualistic. But this process also involves joining, or creating, new communities — communities of unique, self-creating, highly autonomous individuals, but communities nonetheless, and those communities are by definition public.

The truth is, American evangelicals are just as individualistic as American atheists. In terms of their basic postmaterial needs, though obviously not their values, secular self-creators have much in common with evangelical Christians. The opening sentence in Rick Warren's *The Purpose-Driven Life* — "It's not about you" — is often misunderstood as a rejection of individualism. It is not. In fact, the whole book is about God's plan, not for the human race or for the earth but for *you,* the individual.

What makes *secular* self-creators unique is not simply that they have rejected the idea that they are God's chosen people nor that they believe they have some special bond with nature, but rather that they have embraced the difficult journey of self-creation. It is not their *being* but rather their *becoming* that orients them. The pre-political institutions that secular self-creators might join will almost certainly be more network-centric than hierarchical and less dependent upon the dense network of strong ties that characterized social institutions of the industrial age.

And secularists tell their own creation stories that they borrow from the sciences. Biologists, for example, describe humans as beings evolved from the earth. The notion that we emerged from fire and rock and amoebas is no less wondrous than the notion that we were created by an intelligent designer. Why do we assume that it is only the latter and not the former that can inspire us? And while there need be no single story, *some* story about the mystery and majesty of life must be told.

The concepts of belonging and fulfillment can be understood as different metaphors for ways of being in the world. To "be long" speaks to extended social ties — from family to work to community to the world. It has a temporal sense as well. To "be long" is to feel connected far back into the human and nonhuman past and to imagine oneself existing far into the future. To "be full" and to "be filled" describes the world leaning into us — constituting us — as individuals. *Fulfillment* is a lovely word because it is a spatial metaphor that evokes the ways in which we shape and imbue our lives with meaning and purpose.

What's needed today is not the privatization of self-creation, but a willingness to see in the emerging values and new identities of our postindustrial society the power to change the world.

9

PRAGMATISM

I N THE LATE 1960s, Aaron Beck, a young psychiatrist at the University of Pennsylvania, was slowly coming to the conclusion that the psychoanalysis he was doing with his depressed clients simply wasn't working. When Beck looked through the scientific literature, he discovered, to his surprise, that there was little evidence that psychoanalysis had ever worked among people suffering from depression.

Just a few years earlier, Thomas Kuhn had argued in *The Structure of Scientific* that scientific paradigms go into crisis when they fail to explain anomalies. Beck's depressed clients were classic anomalies in the Kuhnian sense. And whether or not one believes that psychiatry and psychoanalysis are scientific, what Beck did next can be fairly described as revolutionary.

Beck abandoned the Freudian view that the key to overcoming mental illness demands dealing with repressed childhood traumas. Instead, he began to look for patterns in the way his depressed patients reasoned. He concluded that depressed people share three core beliefs: *My world is bleak, I'm no good,* and *My future is hopeless.*

The treatment Beck developed involved helping depressed people catch themselves when they made these overgeneralizations, write them down, and then think up alternative stories to redefine the event that had triggered the reactions. Beck's approach worked. His clients were able to break the vicious thought-feeling cycle. After this,

Beck added increasingly challenging activities meant to create a new cycle, this one virtuous. What would later be known as cognitive therapy was born.[1]

Cognitive therapy works because it helps patients re-narrate their lives and in so doing imagine new possibilities and futures for themselves. Our brains are constantly organizing and reorganizing our experiences into narratives that we use to make sense of our lives and our worlds. We remember some experiences, forget others, change the order in which we remember events happening, and conflate experiences that happened at different times and places into one.

In the case of the clinically depressed, it is not that they are remembering or organizing the facts incorrectly. The problem is that depressed patients organize the facts into narratives that become obstacles to living happy, fulfilling, and rewarding lives.

In this way, cognitive therapy is a pragmatic approach to helping people overcome depression. It does not seek to help the patient find the root cause to her unhappiness and eliminate it; it simply helps her reorganize the facts of her life into a narrative that allows her to become the person she wants to be — emphasizing some memories and deemphasizing others, finding new meaning in old experiences, identifying old patterns of thought that result in the old destructive conclusions about herself, and learning new stories through which to understand not only old experiences but new ones too.

In its recognition of the multiplicity of stories that might be constructed from the same event, its rejection of the idea that any of us have an essential being or identity that constrains our possibilities, and its pragmatic application of that insight to the construction of new dreams, new identities, and new futures, cognitive therapy is also a parable for postenvironmentalism.

1.

Modern American liberalism was born, in part, from the philosophical tradition known as pragmatism, which was founded by William

James, Charles Sanders Peirce, and Oliver Wendell Holmes Jr. in the nineteenth century and further elaborated by John Dewey and Richard Rorty in the twentieth. Pragmatists view beliefs as *tools* for shaping reality rather than *mirrors* for reflecting it. President Franklin Roosevelt's experimentalism during the New Deal, which broke from older ideas about the economy and government, can be seen as an outgrowth of pragmatism. It was also during this time that Dewey cofounded and helped lead the American Civil Liberties Union, the American Association of University Professors, the New York Teachers Union, and the New School for Social Research. "It is scarcely an exaggeration to say," the historian Henry Steele Commager quipped, "that for a generation no major issue was clarified until Dewey had spoken."[2] Dewey's view was that ideas and institutions should be contingent — appropriate to the times, not unchanging representations of essential objective interests and values.

While the challenges that America faces and our aspirations for the future are profoundly different than they were in 1930 or 1960, the stories we tell and the institutions we have inherited have largely failed to evolve with the changing times. As we have seen, this is largely due to the fact that the great liberal political accomplishments of the sixties and seventies felt, to those who were part of them, and even to many of us looking back, as if they were the inevitable result of universal principles of truth, justice, and equality prevailing over ignorance, prejudice, and traditionalism.

This view held even as the postwar liberal consensus fractured, and liberalism lost its grip on American politics and government. Liberals forgot the social context that had made those accomplishments possible and instead imagined that something had become broken in politics and government. If science was not politicized, politics not corrupted by special-interest money, and voters not deceived about their true material and ecological self-interest, the truth would again prevail.

But as we have seen, this idea of social change has failed again and again to grapple with the ways that American society has changed. Truths that seemed self-evident forty years ago don't appear to be true to too many Americans today. Americans, both more prosperous and more insecure today, don't define their self-interest in the same way they did in 1970, nor do they understand race relations forty years after the abolition of Jim Crow in the same way that they did when America was still in its grip. And they don't think about global warming in the same way that they thought about the Cuyahoga River catching fire.

Yesterday's narratives, be they of economic inequity, racial discrimination, or belching smokestacks, fail to organize the world into a story of the present that can take us where we need to go. That is not due to a failure of the political system or of government, but of the liberal imagination. The truths of the postwar era, today enshrined in the birth narratives of social movements born of postwar prosperity and enforced by the liberal interest groups that those movements became, have stifled innovation, imagination, and the creation of a new pragmatic politics for more than a generation.

2.

This book is an argument against the politics of essentialism and for a politics of pragmatism. By essentialism we mean thought that reduces complex and multiple realities to a single essence. Essentialism, in this way, constitutes a *metaphysics of stasis,* which imagines that all things have an essential unchanging nature that can be represented objectively. The essentialist imagines that these essences represent the totality of the reality in question and are not dependent upon one's perspective.

By contrast, pragmatist thought begins from the premise that all knowledge is perspectival and all realities are constantly in the process of changing and becoming something else. In this way, pragmatism constitutes a *metaphysics of becoming.* We argue for a metaphysics of becoming rather than a metaphysics of stasis not because we believe the latter is a more objective representation of nature but because we believe it is a more useful tool for describing, shaping, and adapting to the world.

We have seen how Beck's depressed patients reduced their worlds to being bleak, seeing themselves as no good, and their futures as hopeless. And we have seen how environmentalists oversimplify the nonhuman world as harmonious and humans as a reactive primate species. A pragmatic, anti-essentialist politics recognizes that the stories we choose to tell about ourselves and the world have profound implications for the kind of world we create and should be judged by their ability to help us imagine and create the world we want.

Ecology has the potential to be pragmatic if it is contrasted with essentialist thought. And indeed, it was only a matter of time before the scientific eco-logic of interconnected, interdependent *becoming* challenged the essentialism of environmentalism. The ecological study of nonhuman systems should aim not to reduce our manifold, overdetermined, and complex realities to single essences or causes but to explore the interconnections between all things, human and nonhuman, living and dead. Ecosystems are constantly changing. There is no center or essence of any given ecosystem, nor a single meaning of the ways in which it is changing or evolving.

The continuing insistence by environmentalists that reducing carbon emissions is the only rational response to global warming reduces the multiplicity of global warming's possible meanings to humankind's intrusion (pollution) on nature. But there is no single meaning of global warming. Does global warming mean (a) *we're all gonna die!;* (b) we'll be growing bigger and sweeter tomatoes in northern California; (c) humans will survive but find themselves living like prehistoric cave dwellers; (d) we are being punished for our sins against nature; (e) we need better light bulbs and hybrid cars; (f) we must unite the human race around a vision for a clean-energy future; (g) *finally we can build those nuclear power plants we've always wanted!;* (h) we need a cap-and-trade system for carbon emissions;

(i) we must prepare for the worst and hope for the best; (j) the Rapture is on its way; (k) none of the above; or (l) all of the above?

Environmentalists reduce climate science to a single, essential meaning (our greenhouse gas emissions are warming the earth) and an obvious solution (we need to reduce our greenhouse gas emissions). This reduction of a complex phenomenon limits our policy options and, as important, prevents us from fully understanding the problems at hand.

Even if humans had stopped emitting greenhouse gases starting in 1988, when NASA scientist James Hansen announced to Congress that global warming had arrived, all of the changes today resulting from global warming — the melting of Greenland's ice sheet, the slowing of the North Atlantic Gulf Stream, warmer ocean surfaces, and more intense hurricanes — would still be under way. There is so much carbon dioxide and other greenhouse gases in the atmosphere that even if humans stopped emitting new greenhouse gases tomorrow, the planet would continue to heat up for several more decades and probably longer. As surely as the science of climatology tells us that the warming of the earth is caused by humans, it also tells us that a dramatically warmer and transformed climate is almost certainly inevitable.

Environmental leaders claim to be simply representing the climate science, and yet they have consistently downplayed or ignored this aspect of climate science because they did not find it useful to advance the politics, policies, and meanings of global warming that they believed to be most important. They have done so out of the belief that acknowledging the inevitability of both future warming and future impacts would undermine their case for prevention. The reticence among environmental leaders and scientists to acknowledge the irrevocable reality of climate change gives the lie to the presumption that addressing the climate crisis is as simple as dispassionately and objectively translating the science of climate change into public policies.

The relevant question is not "What does global warming mean?"

but "*Which* of global warming's meanings should we elevate into a pragmatic politics?"

3.

To start, let's consider what we know *doesn't* work about the present story environmentalists tell about global warming and put that knowledge to work to create a story that does. We know from extensive psychological research that presenting frightening disaster scenarios provokes fatalism, paralysis, and/or individualistic thoughts of adaptation, not empowerment, hope, creativity, and collective action. We also know that people uniformly describe themselves as happier when they are exercising control over their lives and feel discouraged and depressed when they lose control.

So first we need a story and a plan that makes people feel more in control of their future and better able to address the climate crisis.

We also know that people respond most strongly to threats that they have a mental image of and to threats that involve immediate changes in their perceptions of the world. So second, we need a story that offers immediate, perceptible impacts that can be observed and directly addressed in the present, not the future.

Finally, we know from decades of social psychological research that we are less likely to acknowledge a threat that makes us feel guilty than one that does not, and that we are less likely to acknowledge a threat or take it very seriously if addressing the threat is tied to actions that we perceive as unpalatable — such as radically downgrading one's quality of life.

In sum, a pragmatic politics needs a definition of global warming that is easy to represent by a mental image; communicates an abrupt, perceptible, immediate change in the environment; does not make people feel guilty; has a solution that is not perceived to require tremendous sacrifice; and gives people a sense of control.

In 2005 and 2006, aiming to define global warming in this way, we created a proposal for Global Warming Preparedness, which recasts global warming in terms of preparation for natural disasters and extreme weather. It fulfills all of the criteria listed above and is one of the best opportunities we have for winning both preparatory *and* preventive action on global warming.

Natural disasters are something we are all familiar with and thus can easily form a mental image of. They allow us to respond to well-defined problems and focus on clear and doable solutions. Properly preparing for disasters, and responding to them effectively, enhances one's self-image and sense of control.

When we embrace global warming preparedness we make a concrete commitment to supporting action on global warming. It is a commitment that doesn't require that people feel guilty. But it is a commitment nonetheless, one that can, over time, grow into a much more expansive agenda.

This preparedness approach goes counter to what environmentalists have long held to be true. For more than a decade, environmentalists have attacked any proposal suggesting that adaptation and preparedness might be appropriate responses to mounting evidence that the impacts of global warming have already arrived.

In his 1992 bestseller *Earth in the Balance,* Al Gore wrote, "Our willingness to adapt is an important part of the underlying problem . . . Believing that we can adapt to just about anything is ultimately a kind of laziness, an arrogant faith in our ability to react in time to save our own skin." In response to the EPA's 2002 report on global warming, one Greenpeace executive told the *New York Times,* "Adaptation is like the 'wear sunglasses and a hat' theory of fighting ozone depletion." And in his response to our 2004 essay, the Sierra Club's Carl Pope equated preparedness with "accommodation," and argued that it was fruitless anyway: "We simply won't build a seawall around Florida, much less around the Gangetic Delta in Bangladesh — and a seawall won't stop a hurricane, or save coral reefs."[3]

Environmentalists have attacked adaptation and preparedness

in the belief that taking steps to prepare for global warming — for instance, by building higher seawalls and levees or identifying new water supplies for regions likely to be affected by drought — would undermine their arguments for carbon reductions.

In this, global warming preparedness runs headlong into two rationalist fallacies. The first fallacy is that if people believe that they can adapt to or prepare for the impacts of global warming, they will not be willing to make the sacrifices necessary to avoid it. The second is that if humans don't believe they are the *only* cause of global warming, they will not be willing to take action to reduce or eliminate the human activities contributing to the crisis.

Global warming preparedness offers an alternative to environmentalists who have pegged their strategy on the idea that human reasoning must be linear, literal, and logical. It is not necessary to believe that climate change is 100 percent human caused in order to advocate cutting carbon levels in the atmosphere. Nor does preparing for climate change undermine efforts to reduce carbon emissions. In this context, it is far more likely that people will believe that if we are contributing to global warming — no matter if it's wholly or in part — we should do what we can to prevent the problem from worsening. Once people have accepted that global warming is an immediate threat worth preparing for, the discussion of wider action can resume. When people embrace global warming preparedness, they have made a commitment to one kind of action on global warming. This commitment is the foundation for future action.

In early 2007, the United Nations International Panel on Climate Change recommended an aggressive strategy for adaptation and preparedness, including the need to build higher seawalls. To their credit, environmentalists are beginning to acknowledge that a significant amount of global warming is inevitable and that we need to start planning, adapting, and preparing for it now. And to his credit, the Sierra Club's Carl Pope has reconsidered his dismissal of the efficacy of efforts to prepare for the impacts of global warming.[4]

No agenda to address the crisis can today be considered either serious or science based if it does not have a major global warming preparedness component. Yet to date, no large environmental group has taken up the challenge of campaigning for preparedness.[5]

4.

Special interests are born when people reduce complex reality to single essences and then claim to represent those essences politically. As environmentalism has evolved from a social movement to a special interest, environmentalists have come to see themselves as literally *representing* nature and protecting it from humanity. In politics, environmental leaders and advocates imagine themselves as representing nature or the environment in the same way that organized labor represents union members, the Chamber of Commerce represents businesses, and the NAACP represents African Americans.

Interest-based advocacy groups of all kinds generally define their interests narrowly; once an interest is defined, the group defends it tenaciously. It is for this reason that environmental advocacy and policy development, particularly at the national level, tends to be so reductive, small-bore, and timid. Interest-based advocacy tends to be both loss averse and single-minded. First, hold on to what you've got. Then look to work the legislative, legal, or regulatory process in search of small advantages to further your interest.

This approach tends to work well for powerful economic interests determined to expand their grip on a particular sector of the economy. At various times in the past, it has helped organized labor and civil rights and women's groups in making incremental progress toward the goals of greater equality for their constituents. And it has been a reasonably successful strategy for environmental groups looking to make modest and incremental improvements to existing environmental laws, like the Clean Air Act and the Clean Water Act. But it is an approach that almost always falls short when more dramatic social, political, or economic transformations are in order.

There are few better examples of how this interest-based approach has sabotaged action on global warming than the twenty years of failure to improve fuel-efficiency standards for cars, trucks, and SUVs.

Over the past twenty years, as automobile technologies have improved exponentially, *overall fuel-economy rates have gone down, not up.* The Corporate Average Fuel Economy legislation, or CAFE, was crafted in 1975 more as a way to save the American auto industry, not the environment. With fuel prices rising dramatically after the oil shocks of the early 1970s, and U.S. automakers ill prepared to offer the fuel-efficient vehicles that the public was suddenly demanding, U.S. automakers, the UAW, and environmentalists alike agreed that a modest fuel-efficiency requirement made good sense and would help Detroit more effectively compete with foreign competitors, would reduce dependence on foreign oil, and would reduce air pollution.

The original CAFE regulations worked: automobile fuel efficiency improved dramatically through the late 1970s and early 1980s. But as oil prices eased in the mid-1980s and the gains from relatively easy efficiency improvements were realized, CAFE's efficiency gains began to stall, and, with plummeting oil prices in the mid-1990s and the rise of the SUV, those gains were reversed. By the early 1990s, environmentalists resolved to raise fuel-efficiency standards further, this time without the cooperation and support of the UAW or the auto industry.

From this point on, CAFE would become an environmental proposal and would enter a special-interest purgatory — prioritized by environmentalists and bitterly opposed by both the UAW and the auto industry — in which it has languished ever since. In turning CAFE into an environmental issue prosecuted solely by the major national environmental groups, environmentalists came to concern themselves with the automobile only insofar as gasoline goes into it and pollution comes out of it, and they failed to address any of the larger dynamics that were driving the failure of American automakers to produce competitive, efficient, and innovative automobiles.

They transformed CAFE from a win-win proposition that had helped both to get the American auto industry back on its feet and to reduce pollution in the 1970s and 1980s into a zero-sum game in which either the environment, represented by environmentalists, or the auto industry and auto workers, represented by the UAW and the Big Three, would win.

The conventional wisdom today is that the auto industry and the UAW "won" the CAFE fight. And in the literal, interest-group view of the fight, they did. The auto industry and the UAW avoided what they considered to be onerous regulations that would marginally reduce profitability. And the environmental groups failed to win further regulatory increases in fuel efficiency.

But in the larger sense, the conventional wisdom is all wrong. Today the American auto industry is in a state of gradual collapse, in no small part because auto executives failed to innovate. Japanese automakers are eating away at the American market share with cleaner, more efficient, and outright *better* vehicles. And the American auto industry has hemorrhaged hundreds of thousands of high-paying union jobs as the Big Three shuttered factories and moved much of their production overseas, where labor is cheaper. In this sense, none of the special interests have done especially well in representing their interests, and certainly none could be said to have won.

Like so many issues we have examined in this book, from inner-city asthma to deforestation to global warming, the multiple causes of the auto industry's failure are complex and overdetermined. To name just a few, they include an aging, antiquated, and undercapitalized industrial base; an aging work force; enormous legacy health and retirement costs; and deeply encrusted cultures among management and labor that stifle both innovation and flexibility.

And yet, for two decades environmentalists have failed to imagine a political strategy or policy framework that might address fuel efficiency in the context of the larger crisis faced by the American auto industry. Indeed, environmentalists defined all of that context as out of their field of concern. When we interviewed the environmental movement's most prominent and influential champion for CAFE, the Sierra Club's Dan Becker, he was emphatic on this point.

> We need to remember that we're the environmental movement and that our job is to protect the environment. If we stray from that, we risk losing our focus, and there's no one else to protect the environment if we don't do it. We're not a union or the Labor Department. Our job is to protect the environment, not to create an industrial policy for the United States. That doesn't mean we don't care about protecting workers.[6]

The problem is that a new industrial policy is precisely what the United States needs to make the transition to a clean-energy economy.

5.

The high cost of health care for its employees and pensioners is one of the biggest crises the U.S. auto industry faces. Because Japan has national health care, its auto companies aren't stuck with the bill for its retirees, freeing up their resources to invest in things like research and development. Even when Japanese companies produce cars in the United States, they benefit from a far younger work force whose health costs are lower. Yet this issue has long been treated by environmentalists as something that has nothing to do with CAFE or the set of concerns they define as environmental.

Instead of attempting to understand how the industry and its unions reason about their future and then figure out how fuel-efficiency improvements might be presented as consistent with — rather than irrelevant to — the industry's own estimations of its problems, environmentalists have tried to sprinkle a little sugar on top, in the form of a few new tax credits, to make the medicine of fuel-efficiency regulation go down.

In his response to our essay, the Sierra Club's Carl Pope wrote, "Environmentalists have also continuously and intensely explored ways to make the program work for both the unions and the domestic manufacturers by offering tax credits or other mechanisms to finance the necessary catch-up by Detroit."

Pope was telling the truth. He and other environmentalists did indeed repeatedly offer to lobby for tax credits in exchange for CAFE. The problem is that tax credits for higher fuel-economy standards weren't what the automakers wanted or needed. It was inconceivable to Pope and other environmentalists that they might achieve their goal if they concerned themselves with problems such as health care.

In mid-2005, with GM and Ford teetering perilously close to bankruptcy and losing tens of thousands of jobs and with pressure growing for Congress to bail out Detroit, we suggested a different approach and proposed in a white paper that any investment the federal government makes in the auto industry be tied to increases in vehicle fuel economy. The Health Care for Hybrids proposal, which we created together with Senator Barack Obama, Representative Jay Inslee, and the Center for American Progress, would link fuel-economy increases to relieving health care costs for automakers.

Our argument was straightforward: Congress may be asked once again to bail out Detroit. After 9/11, Congress bailed out the airline industry and demanded nothing in return. This time, Congress should demand accountability. The proposal recognizes that American automakers are burdened with massive health care costs that their foreign competitors, blessed with national health care systems, don't have.

This competitive disadvantage partly explains Detroit's failure to invest adequately in research and development and thus produce better vehicles. But Detroit is not free from blame: the Big Three have stubbornly refused to innovate and are now stuck selling gas guzzlers that fewer Americans want. What's needed, we suggested, is a new deal for the auto industry — one that is entirely voluntary. If automakers want help covering the health care costs of their workers, under the Health Care for Hybrids Act they must increase the efficiency of their vehicles.

Having defined their job as *representing* nature, environmentalists never even considered that ecological opportunity might lie within Detroit's shrinking profits, declining markets, and health care crisis. That's because autoworkers, industrial plants, health care, and pensions aren't part of nature in the essential and reductive way that environmentalists think about it. To their credit, the Natural Resources Defense Council endorsed Health Care for Hybrids, and in this, as with much that has occurred over the past several years, there is great promise.

Ecological thinking at its best sees complex relationships and interactions among industrial plants, workers, health care, energy, and pollution, and economic growth as no less a part of the planet's ecosystem, and no less natural, than a rain forest. A healthy, innovative American auto industry deeply invested in producing cutting-edge, highly efficient automobiles is essential to America's clean-energy future.

6.

The problem with today's liberal interest-group categories is not that they exist: none of us would be capable of thinking or acting if we did not categorize the world in one way or another. The problem is that those categories have come to be seen as essential and unchanging reflections of reality rather than as conceptual tools — which is to say, *ideological constructs*. Those constructs were once very useful in advancing a progressive politics, but today they get in the way as often as they help us.

Environmentalists and liberals claim to be the intellectual heirs to Darwin, but most refuse to accept the fact that their own beliefs, sciences, and truths are adaptations to past realities — not reflections of them. Science has never been a reflection of nature, much less the only vehicle for the expression of the truth. There have always been multiple, contradictory, and overlapping sciences, truths, and natures.

No sooner does a physicist — a practitioner of the supposedly hardest of the hard sciences — find the smallest thing in the universe (atomic particles) than another physicist comes along to announce something smaller (vibrating strings). Biology gravitates toward essentialism no less than physics gravitates toward final, unified explanations. Most practicing biologists believe that species are "natural kinds," existing out there and objectively in nature, independently of us humans. But what biologists call natural kinds are always already human categories. Consider the fact that there is no single meaning of the word *species* that covers all the ways biologists define the category *species* in practice. It is we humans, after all, who draw the boundaries between species.[8]

If an honest appraisal of the sciences of biology and physics forces us to acknowledge that the categories those sciences depend on are constructed by humans, not found in nature, is it really so hard to acknowledge that liberal interest-group categories are constructed as well and, as such, might better serve the needs of this political moment if they were reconstructed and reimagined? The mass of progressives working at liberal interest groups behave as though their issue categories exist separately from the rest of the world. Reproductive rights advocates see the creation of a new social contract as liberal foreign-policy experts see agricultural subsidies and as environmentalists see a new industrial policy: "not my issue."

In a series of lectures he delivered in 1997, the philosopher Richard Rorty attacked the academic left's obsession with philosophy as a kind of solipsism, utterly disengaged from the real world of politics. But perhaps the problem is the opposite; namely, the anti-philosophical turn that the political left took in the 1970s and 1980s.[7]

Rorty argued that the left should "kick its philosophy habit," ignoring the fact that the most influential figures in the left today do not reside within the ivy fortress of the academy but rather in liberal interest groups. It is interest-group leaders, not academic postmodernists, who determine the Democratic Party's agenda on virtually everything: global ecological crises, foreign policy, reproductive rights, agriculture, labor, and a new industrial policy. The truth is that the left in general and the Democratic Party in particular would benefit from a bit of the deconstruction, and concomitant reconstruction, that Rorty's powerfully pragmatic philosophy can provide.

In this sense, Rorty had his guns pointed in the wrong direction. It's not that the academic left needs to engage in the real world of politics. Rather, the political left needs to get a philosophy habit.

7.

In the West, nature has been seen at various times as something to either fear or love, as something chaotic or harmonious, and as superior or inferior to humans. But what has remained dominant is the belief that something called nature exists, that it has an essential quality separate from human beings, and that it can be understood objectively through science — *as a thing*.

Environmentalists have long disparaged human power and adaptability because they believe that nature has a metaphysical essence — a single meaning or mode of being — that cannot or should not be changed, violated, or intruded upon.

Rachel Carson denounced humankind's self-destructive use of chemicals against nonhuman natures as a consequence of our will to control nature. Al Gore and others insist that adaptation is not an option in dealing with global warming, because they make no distinction between changing nature and destroying nature. If nature has a single, essential, static meaning, then to change nature is by definition to destroy it.

Conservatives have their own theological construct that must

not be violated: the market. The market, for conservatives, like nature, for environmentalists, is *a thing* separate, sacred, and inviolable.
Indeed, it is natural, born of human nature and iron economic law.
Sins against the market, like sins against nature, will be punished. Environmentalists argue that societies that violate nature's laws will ultimately collapse, and market fundamentalists argue that economies
that violate the laws of the market will ultimately collapse.

We often think market fundamentalists and environmentalists
are different largely because they set the terms of political right and
left. But both essentialize their respective objects of veneration as
harmonious, in balance, and *rational.* Environmentalists believe that
nature tells them what to do through science, whereas market fundamentalists believe that the market tells them what to do through economics. Environmentalists imagine that they are protecting nature
from people, just as market fundamentalists imagine they are protecting the market from people (that is, from the government). When
there are crises in nature, the market, or both, environmentalists
and market fundamentalists tend to blame irrational human intrusion.

We speak of the market in the singular, but markets, like nature,
have no spirit, essence, or invisible hand. They are little more than (to
borrow from the textbook definition) mechanisms for efficiently dis-
tributing scarce resources. Markets have no universal end goal, nor
do they operate according to a universal set of values. Some markets
are good, which is to say that they serve our values, and others are
bad, which is to say that they don't.

Many on the left say that some things, such as health care, access
to water, and energy, should a priori never be traded on the market,
while many on the right say that all things should be traded on the
market, because the market is good. Both kinds of generalizations are
equally inane. Slavery wasn't bad merely because slaves were bought
and sold on the market. Slavery was bad because condemning a fellow human being to servitude against his or her will is immoral. The

questions to ask are thus "When are markets useful and when aren't they?" and "What values and future should markets serve?"

Market fundamentalists claim that human governance is always an impediment to markets, but in fact human governance is what makes markets possible. Government provides the courts, regulates commerce, and sets the rules for society. The ban on murder can be seen as an impediment of the free market. By prohibiting the murder of one's competitors, one is restricting commerce. While it's true that market fundamentalists don't argue against prohibitions on murder (at least to our knowledge), neither do they fully acknowledge the ways in which all human laws — our codified values — *create* the conditions for that particular domain of human behavior known as the marketplace.

All markets are constructed and shaped by humans through laws and regulations as well as through values. The idea that one should or can hold a political position on markets in general is absurd, and yet left and right, environmentalists and market fundamentalists alike, continue to argue in the abstract about whether they are pro-business and pro-market.

Incredibly enough, there are still people who call themselves progressive and imagine themselves to be anticorporate and antiglobalization. They argue with people who imagine themselves to be proglobalization or, more nonsensically, free traders. To declare oneself as proglobalization or as a free trader is as meaningless as declaring oneself antiglobalization, anticorporate, or in favor of fair trade. The vast majority of environmental and other NGO professionals, in fact, wage campaigns to encourage corporations to improve their social and environmental performance, not eliminate markets or corporations.

The endless debates between environmentalists and market fundamentalists ultimately pit one theological construct, nature, against another, the market. For either camp to expect that reason or science or economics will prevail in such a debate is folly. Just as humans, from indigenous tribes in the Amazon to Americans in Yosemite, are always constructing natures, whether for development or conservation or something else, humans are always constructing markets, whether for positive ends, like ecological restoration, or negative ends, like slave trading. Once we abandon the belief that there exists a nature or a market separate from humans, we can start to think about creating natures and markets to serve the kind of world we want and the kind of species we want to become.

8.

Today, the effort to save the Amazon is being undermined by the quasi-religious adherence to both nature and the market. In imagining that they are trying to protect nature from humankind, conservationists define virtually all of the key drivers of deforestation — Brazil's debt, its lack of land reform, and its need for economic growth — out of their politics.

Imagine that in 1964 Brazil's democratic government had been invaded by a foreign government that then ruled as a military dictatorship for the next quarter century. Seeking to quell internal dissent, the illegal government borrowed billions of dollars from foreign banks and governments to destroy one of Brazil's national treasures, the Amazon forest. Forty years later, Brazil became democratic again but still found itself paying back the debt incurred by an illegal, foreign government. It did so by starving its people and further destroying the Amazon, constantly deferring its superpower potential.

Everything in the above paragraph is true except for one fact: the illegal government was not foreign. But Brazil's dictators borrowed from foreign governments and was happily supported by the United States. The dictatorship tortured, imprisoned, and murdered Brazilians who demanded freedom and democracy. The dictatorship's debts paid for gargantuan colonization schemes in pristine forest that resulted in hardship and violent barbarities. The debts are today being serviced thanks to the country's cattle and soy boom — a boom built atop the ashes of ancient forest riches.

Brazil's dictatorship debt, while freely traded everywhere from Tokyo to Frankfurt and owned by European mutual-fund holders and middle-class Brazilians alike, is a kind of postmodern peonage. What is being paid back is not a debt freely incurred by democratically elected representatives of Brazilian citizens in order to grow their way out of hunger and misery, but rather debt that an illegal dictatorship created against the free will of the Brazilian people. It is cruel, counterproductive, and inhumane to force a Brazilian mother to pay back debts incurred by corrupt generals before she was born. That her dictatorship debt payments come at the expense of her children — children who must wander the streets of Rio de Janeiro in search of meager sustenance — is a moral abomination. Why should the citizens of a democratic nation be forced to pay the debts incurred by an illegal government?

For more than four decades, ordinary Brazilians have been told that if they struggle and suffer now, greatness will eventually come. But Brazil can neither overcome its material deprivation nor achieve greatness as long as it must give up food for its poorest children and destroy one of its most valuable national assets in order to make payments on a debt that it has already paid off many times over. Brazil's debt is a central, if not *the* central, obstacle to Brazil's achieving prosperity. Every cent Brazil pays each year to service its debt is one less cent that the country has to invest in the research and development, education and infrastructure, that could make Brazil an economic superpower rather than merely an agricultural one. In accepting that Brazil's debt is an obligation that must be repaid, traded for land, or forgiven, environmentalists and progressives alike pay obeisance to the market even as they rail against it.

9.

Market fundamentalists believe that there is nothing wrong in requiring poor Brazilians to pay back a loan that an illegal government took out more than thirty years ago to destroy the Amazon. They believe that there is a law higher than democracy or freedom or justice: the law of the market. And they think that once something is traded freely in markets, it is holy and good. Once the dictatorship debt is freely traded, it is magically cleansed. What could possibly be immoral about something one can freely buy through Charles Schwab?

And yet, throughout human history, all sorts of terrible things have been freely traded — from slaves to tropical hardwoods to dictatorship debt. Today, if you want to buy dictatorship debt, you can do so. Most who own dictatorship debt, including Brazilians themselves, do not know of its connection to the hunger of Rio's street children or the destruction of the greatest forest in the world. Market fundamentalism depends on masking the ways in which markets *always* rely on government and human values.

Market fundamentalists will insist that erasing Brazil's dictatorship debt is a human intrusion into the natural functioning of markets. But because those debts were incurred illegally, the dictatorship debts must not be respected. Getting rid of them and helping Brazil grow economically offers hope both for ridding Brazil of the scourge of hunger and halting the decimation of the Amazon.

In calling for the elimination of the dictatorship debt, we are most definitively not calling for debt forgiveness, a concept that implies that the debtor countries committed some sin for which they should be forgiven. It is not the indebted people of Brazil who should be forgiven but those who blithely insist that the dictatorship's debt is either moral or legal. Brazil doesn't need our forgiveness. It needs justice to be served.

10.

There is no shortage of liberal-minded Americans who continue to dream Rousseau's utopian dream of a time when we will all unite around our common interests, and conflict in politics will die away. You know an environmentalist holds this dark fantasy when he says,

"Global warming should be above politics." But other than a meta-physical dictatorship, what could possibly be above politics? Anyone who wants some arbiter that is above politics is either appealing to a theological authority or misunderstands what is meant by *politics*.

If politics is our self-governance as a species, then it is the highest form of collective authority there is. The truth of the collective is that it is multiple, contradictory, and divisive. There is no single public interest. To deny the multiplicity, as many neo-Rousseauians do, is to miss something fundamental to politics. Politics is about making decisions.

Given what's at stake and the quickly diminishing time frame for action, we must quickly embrace a politics that understands that humans are constantly creating new natures. We must abandon the environmentalism that thinks of itself as representing and defending — but never imagining, constituting, or creating — nature. What environmentalists must do now is recognize that all politics is constitutive, in the sense that politics is always an act of creation. Overcoming the ecological crises and realizing humankind's potential will require abandoning efforts to return to some Edenic past. It will require going through modernity and materialism, not avoiding them. And it will require that we leave environmentalism behind as we construct a postnatural politics.

In 1969, one of the founding fathers of American environmentalism, René Dubos, called for environmentalism to become a new religion, one capable of healing humankind's relationship with nature. But what we need is not a new *religion*, especially not one grounded in a theological view of harmonious nature. What we need is a new *politics*, one capable of fully appreciating humans as evolving beings filled with messy contradictions, emotions, and rationalities who are constantly adapting to and transforming their environments and their realities.

What was special about environmentalism —— its commitment to ecological thinking, its scientific questioning, and its understanding of humans as beings of earth — must be brought forward into the new politics. Its monotheism, its philosophy of limits, and its irrational rationalism must be left behind.

As the earth warms, forests disappear, and the Arctic melts into the oceans, new natures will emerge all over; it will become increasingly untenable for anyone to claim to represent some essential nature or environment. We will find ourselves openly imagining, constructing, and fighting for particular natures and environments amenable to human civilization. Just as modernity has replaced the question "Who are we?" with "Who shall we become?," the ecological crises will replace the reductionist question "What must we do to save the environment?" with "What new environments can we imagine and create?"

Pluralizing singular categories is a simple way to free ourselves from essentialism. In abandoning Nature, we can embrace the multiplicity of human and nonhuman natures. In abandoning Science, we can embrace the various kinds and practices of the sciences. In leaving behind the belief in a single objective Reason, we can better understand that we have multiple ways of reasoning about the world. In rejecting an essentialist view of the Market, we can embrace the power of markets to achieve our social and ecological goals.[9] If we are to imagine a new future for humans and nonhumans we must abandon essentialist politics and construct a politics of possibility and potential.

Environmental leaders, like psychologist Aaron Beck's depressed patients, tell the same reductive story about global warming over and over again because they believe there is no other story to tell. But that doesn't mean we cannot create a new, pragmatic politics based on the notion that the world is wondrous and beautiful, not fallen. The ecological crises are not our fault — we were born into them. Nor were they the fault of past generations — they were doing what they could to create a better future for us. And the future is hardly hopeless. Humans are the most powerful, creative, and adaptive species ever to roam this remarkable planet. We have overcome hunger, disease, and oppression — we can overcome ecological crises.

When we called on environmentalists to stop giving the "I have a nightmare" speech, we did not mean that we should close our eyes to our increasingly hot planet, the destruction of the Amazon, or continued human suffering. Rather, we meant that we all should open our eyes to the multiplicity of ways we can see and experience the world. Global warming could bring drought, disease, and war — *and* it could bring prosperity, cooperation, and freedom. The future is not destined to be dark or bright, fallen or triumphant. Rather, the future is *open*.

10

GREATNESS

S HORTLY BEFORE THE 2004 election, the journalist Ron Suskind wrote an article for the *New York Times Magazine* about President Bush's faith-based presidency. In it, he quoted an unnamed senior Bush administration adviser contemptuously referring to Suskind as part of the Washington, D.C., "reality-based community."

> The aide said that guys like me were "in what we call the reality-based community," which he defined as people who "believe that solutions emerge from your judicious study of discernible reality." I nodded and murmured something about enlightenment principles and empiricism. He cut me off. "That's not the way the world really works anymore," he continued. "We're an empire now, and when we act, we create our own reality. And while you're studying that reality — judiciously, as you will — we'll act again, creating other new realities, which you can study too, and that's how things will sort out. We're history's actors . . . and you, all of you, will be left to just study what we do."[1]

Liberals were, needless to say, aghast. But they were also delighted: the aide's statement confirmed for them that the Bush administration was little more than a gaggle of deluded ideologues. Every time a liberal columnist or blogger wanted to make this or that point about the war or Hurricane Katrina or global warming she would often begin by smugly declaring herself a proud member of the "reality-based community."

But the reality-based community's reaction spoke as loudly to the failure of the liberal imagination as it did to the Bush administration's delusional foreign policy. What the senior Bush aide had said was, in fact, dead on: in Iraq, the Bush administration has created a new reality — and liberals have done little more than study it.

Few things today are more offensive to liberal and environmental leaders than the notion of humans creating new realities, political or otherwise. But a certain hubris is always required of people, be they environmentalists or neoconservatives, who wish to change the world. Ironically, environmentalists and liberals today view hubris as a transgression to be punished, whereas conservatives see it as a point of pride.

Today, "changing the world" better describes the mission of American conservatism than American liberalism. Coming of age in the late sixties and early seventies, when roughly half of all American voters identified themselves as Democrats and just one-fifth as Republicans, conservative strategists were forced to constantly create political realities more conducive to the Republican vision and conservative values. By contrast, most Democratic and liberal leaders have settled for an unimaginative realpolitik that sees all large and transformative visions as dangerous.

The problem is that the realities that thirty years of political dominance have created, from disarray and radicalism in the Middle East to massive budget deficits to rising social insecurity, will in the long term accrue to American conservatives, as long as progressives remain reality based and not imagination based. America's humiliation in Iraq, nuclear proliferation in Iran, and the threat of further terrorism — these are likely to trigger greater insecurity and fear, psychological states that history and a great deal of empirical research show drive conservatism far more strongly than liberalism.[2] In this context, to remain reality based is to accept a status quo that is bad for progressives everywhere from Tehran to Akron.

1.

In 2006, as Iraq spiraled ever further out of the Bush administration's control, the neoconservative intellectual consensus behind the U.S. invasion began to come apart. That year, a leading American neoconservative, Francis Fukuyama, published *America at the Crossroads: Democracy, Power, and the Neoconservative Legacy,* in which he broke from neoconservativism.[3]

Once in 1998 and again in 2001, after September 11, Fukuyama signed open letters urging military action be taken to remove Saddam Hussein from power.[4] "Failure to undertake such an effort," the 2001 letter to President Bush warned, "will constitute an early and perhaps decisive surrender in the war on international terrorism."[5] *America at the Crossroads* served as his mea culpa, albeit one that required that he airbrush his advocacy of U.S. military action in Iraq after 9/11.[6]

In the book, Fukuyama reasserted an argument he had made fourteen years earlier in *The End of History and the Last Man,* which was that the collapse of the Soviet Union and the democratization of former right-wing dictatorships in places like Chile signaled the end of history prophesied by the philosopher Hegel in the early nineteenth century. By *history* Fukuyama meant *History:* the grand evolution of human societies from slavery to dictatorship to democracy. Fukuyama acknowledged that some countries were still dictatorships, and that there were still totalitarians, including Muslim (and other religious) fundamentalists, who saw modernity as decadent and wanted to impose an ancient moral order upon the world. But history as the unfolding of human development, Fukuyama argued, had ended. What today appears as diverse forms of rule are, in fact, countries in different stages of the same journey. "Rather than a thousand shoots blossoming into as many different flowering plants," Fukuyama wrote, "mankind will come to seem like a long wagon train strung out along the road."[7]

It was a strikingly liberal, even quasi-Marxist, argument for a conservative to make.[8] For much of the postwar period, conservatism had been defined around defending tradition, not progressing beyond it. In 1955, in his introduction to the very first issue of the *National Review*, William F. Buckley described his new magazine's mission as follows: "It stands athwart history, yelling Stop, at a time when no one is inclined to do so, or to have much patience with those who so urge it."[9] But arriving first in essay form in 1989 and in book form in 1992, Fukuyama's modernization thesis captured the triumphalist conservative sentiment after the fall of communism better than the older traditionalist one.

In *America at the Crossroads*, Fukuyama argued that the neoconservative error in Iraq came from believing that the organic internal process of national development — of rising prosperity, changing social values, and increasing democracy — could be accelerated through military means.

> What is *initially* universal is not the desire for liberal democracy but rather the desire to live in a modern society, with its technology, high standards of living, health care, and access to the wider world. Economic modernization, when successful, tends to drive demands for political participation by creating a middle class with property to protect, higher levels of education, and greater concern for their recognition as individuals. Liberal democracy is one of the by-products of this modernization process, something that *becomes* a universal aspiration only in the course of historical time.[10]

Today the view that prosperity and changing social values are preconditions of liberal democracy and progressive social movements is rejected by both the neoconservative right and much of the left. The neoconservative right, as Iraq shows, came to believe that democracy could be forcibly imposed from the outside. The environmentalist left, as we have seen, tends to view prosperity as an obstacle to, rather than a prerequisite for, the emergence of ecological concern and environmentalism. And much of the labor movement and the antiglobalization left opposes globalization for destroying old industrial jobs and undermining traditional cultures.

Like Fukuyama, we find modernization theory to be a powerful tool for understanding and shaping the world. Where we diverge from Fukuyama is in our rejection of his reduction of modernization to a single, essential, and inevitable process and in our support for the expansion of, rather than limits to, personal freedom for self-creation in postmaterialist (what he calls posthistorical) countries like the United States.

2.

Fukuyama felt that, of the left-wing and right-wing criticisms of his thesis, the criticisms from the left were easier to answer. The most common progressive objection to the end-of-history argument was that the remaining social injustices in democratic societies create new conflicts that move history forward. In *A Tale of Two Utopias*, the liberal critic Paul Berman criticizes Fukuyama on this point:

> "Today in democratic America," [Fukuyama] writes, "there is a host of people who devote their lives to the total and complete elimination of any vestiges of inequality, making sure that no little girl should have to pay more to have her locks cut than a little boy, that no Boy Scout troop be closed to homosexual scoutmasters, that no building be built without a concrete wheelchair ramp going up to the front door." That's a pretty dismissive way of describing society's campaign for more equality. But with all due respect to the skeptical conservative instinct, let us postulate

that social problems of the sort that are invisible only to conservatives actually exist, and that several grounds other than prissy middle-class hyper-egalitarianism might lead to outbreaks of social indignation and all kinds of social crises in the future. What then?

But this criticism misses the central point of Fukuyama's book that, at least in wealthy, posthistorical countries, outbreaks of indignation and social injustice wouldn't be solved by a new stage of human development, such as socialism, but rather by *more* liberal democracy — a point that Fukuyama's liberal critics, Berman included, largely agreed with.

A far stronger challenge to his end-of-history argument, Fukuyama felt, came from the right, which he located in the nineteenth-century philosopher Friedrich Nietzsche's concept of the "last man." In contemporary terms, the last man is "the ultimate couch potato."[11] Fat, comfortable, and contented, the last man is uninterested in realizing his personal potential. His indifference, his apathy, and his lack of creativity form a kind of nihilism.

Fukuyama sees the last man's affliction as a kind of *isothymia* — the atrophying of *thymos,* our spirited desire for social recognition — that characterizes nations that have achieved unprecedented levels of freedom and prosperity. Traditionally, humans have achieved greatness by striving against material privation and for freedom. Without material privation or oppression against which to struggle, how will humans define life's purpose? How will we achieve greatness?

Fukuyama's solution is for people to re-embrace traditional family values.

Many of the problems of the contemporary American family — the high divorce rate, the lack of parental authority, alienation of children, and so on — arise precisely from the fact that it is approached by its members on strictly liberal grounds. That is, when the obligations of family become more than what the contractor bargained for, he or she seeks to abrogate the terms of the contract.[12]

Fukuyama's complaint is consistent with the concerns raised by the neoconservative Daniel Bell. In 1976, he published *The Cultural Contradictions of Capitalism,* which argued that the prosperity created by capitalism was undermining the Protestant values — such as the work ethic, personal savings, and personal responsibility — that had helped give rise to capitalism in the first place.[13] Bell's scorn for the modern "life style" — and "self-fulfillment" — echoes throughout the pages of *The End of History.*[14]

But where Fukuyama embraces a version of modernization theory, which is a narrative of human rising and overcoming, Bell embraces a narrative of tragedy, of humankind's fall from the Protestant values such as the work ethic.

> When the Protestant ethic was sundered from bourgeois society only the hedonism remained, and the capitalist system lost its transcendental ethics . . . [A] society that fails to provide some set of "ultimate meanings" in its character structure, work, and culture, becomes unsettling to a system.[15]

Or, rather, it became unsettling to Bell and other conservatives. The truth is, as Richard Florida points out in *The Rise of the Creative Class,* the traditional Protestant work ethic has been replaced by a more flexible and more creative relationship to work, employment, and one's identity as a laborer. It was the melding of the more hedonistic bohemianism with the Protestant work ethic that was behind the rise of the highly productive creative class. "They morphed into a new work ethic — the creative ethos — steeped in the cultivation of creativity," Florida wrote.[16]

For his part, Fukuyama sees the solution to the hedonism of posthistorical societies in limiting freedom. "This suggests," he wrote, "that no fundamental strengthening of community life will be possible unless individuals give back certain of their rights to communities, and accept the return of certain historical forms of intolerance."[17]

Fortunately for us, neither Bell nor Fukuyama nor the larger conservative movement can put the genie of new freedoms and new values back in the bottle. Individuals today have the right to end un-happy and cruel marriages. Women have won the right to pursue their happiness, even when it threatens men. Same-sex couples have won the right to love and live with each other without fear of institutional persecution. And consistent with the logic of modernization, same-sex marriages will one day be recognized by societies and states throughout the world.[18]

In the end, Fukuyama's conservatism gets in the way of his progressive view of history. Capital-*H* History may have ended, at least for now, but our social values and our politics will continue to progress. The question is whether the ecological crises we face can be overcome without human civilization's being forced to undergo the kinds of calamities that would signify, in Fukuyama's reading, the restarting of History.

3.

In contrast to Bell and Fukuyama, Nietzsche advocated the end of traditional Judeo-Christian moral values. He believed that the way for humankind to achieve greatness was through the creation of *new* values appropriate for the age.

Nietzsche argued that the decline of traditional religious authority — "the death of god" — and the growing disbelief that there was an external Being controlling human lives were the crises at the heart of modern nihilism. And within those crises lay an opportunity for humans to take responsibility for their own governance by creating new values and by replacing philosophies of resentment, pessimism, and limits with a philosophy of gratitude, overcoming, and possibility.[19]

The problem with Nietzsche, Fukuyama, and Bell is that they either couldn't or wouldn't imagine a human civilization in which *all* of its members have the freedom to pursue their highest potential. Nietzsche believed that creative greatness could be achieved only atop aristocratic societies. Fukuyama and Bell believe that human potential should be constrained by traditional moral values.[20]

A progressive view of history rejects both Fukuyama's and Bell's embrace of traditional moral values and Nietzsche's embrace of aristocracy. The decline of the traditional family and the Protestant work ethic, far from undermining prosperity, have proven to be vital to the rise of a vibrant postindustrial capitalism.

And in affluent postindustrial societies there is no reason why every citizen could not adopt the virtues of strength, pride, and excellence that Nietzsche valorized. Those values are entirely consistent with democratic, egalitarian politics. Indeed, these are increasingly the values of many Americans who hold autonomy, individuality, and self-creation above virtually all else.

Americans today aspire to be unique individuals, to be autonomous and in control of their lives, and to be respected and recognized as such by those around them. They no more aspire to be common than they aspire to be poor. The class wars of the Great Depression and the New Deal are dead and gone, and the politics of the common man with it. Today's common man, with his pension and retirement accounts, more likely than not owns the means of production — an ironic fulfillment of Marx's prophecy. He eats like the aristocrats that Nietzsche exalted and lives in a castle with central heating and a flat-panel television.

Throughout this book we have argued that for people to become compassionate and generous, they must become, and feel, secure, wealthy, and strong. This was the basis of our argument for global warming preparedness, health care reform, and the construction of a postmaterialist politics aimed at meeting our needs for esteem, status, belonging, and purpose.[21] By contrast, environmentalism is hobbled by its resentment of human strength and our desire to control nature, and liberalism by its resentment of wealth and power. The morality of self-creation offers environmentalists and liberals alike the recognition of wealth, power, and self-mastery as virtuous, not evil.

Those of us who are fortunate enough to have met our basic material and postmaterial needs should feel neither guilt nor shame at our wealth, freedom, and privilege, but rather *gratitude*. Whereas guilt drives us to deny our wealth, gratitude inspires us to share it. It is gratitude, not guilt, that will motivate Americans to embrace the aspirations of others to become as wealthy, free, and fortunate as we are.

4.

History is, for Fukuyama, guided by a single *telos*, or end: inevitable progress. Given this plainly deterministic view, one would be justified in believing that the grisly nightmare that is Iraq disproves his hypothesis: if history were inevitably moving toward liberal democracies, what explains liberal America's *illiberal* invasion of Iraq, which has fueled religious totalitarianism?

But Fukuyama never suggested that history would *only* progress, merely that it would generally move in a progressive direction, and on that score the Iraq War has done little to undermine Fukuyama's thesis. Around the world, countries generally are continuing to prosper and move toward greater levels of freedom. And in at least two ways, the carnage in Iraq gives weight to two concerns expressed in *The End of History and the Last Man*.

Toward the end of that book, Fukuyama worried that the peace and prosperity of the early 1990s might create a boredom resulting in a lust for new wars, a kind of *megalothymia* that could send human civilizations back into barbarism and thus restart history. The invasion of Iraq, the election of fundamentalist extremists in Iran and

Palestine, and nuclear proliferation in Iran and North Korea gave weight to Fukuyama's concern. And the failure of the Iraq invasion to result in a stable government also demonstrates the correctness of Fukuyama's argument — and modernization theory — that liberal democracies cannot be imposed from the outside but must be arrived at through an internal process of development.

The greatest challenge to Fukuyama's argument in *The End of History* came neither from chaos in Iraq nor continuing demands for social justice, as Berman would have it, but rather from ecological crises and biotechnology. When it comes to the question of global warming, Fukuyama blinks. In his 1993 book he dismisses ecological concerns altogether. And in the 2006 afterword to *The End of History*, Fukuyama endorses the market fundamentalist's argument that avoiding the worst effects of global warming through making the transition to clean energy simply cannot be achieved without harming the economy.

> If some of the more dire predictions about global warming are correct, it may already be too late to make the sorts of adjustments in hydrocarbon use that will prevent massive climate change, or else the adjustment process will itself be so disruptive that it will kill the economic goose that is laying our technological golden eggs.[22]

Fukuyama rejects outright the idea that we can overcome global warming because to do otherwise would challenge the foundation of his *End of History* argument. To understand why this is the case, we must first understand that for Fukuyama, supposedly following Hegel, what drives history forward is something called "the Mechanism." Simply put, the Mechanism is humankind's domination of nature through science and technology. The end of history would arrive at the moment humans learn how to completely dominate nature.

The antienvironmentalist Fukuyama, like the environmentalists he parts ways with on global warming, is at bottom a rationalist who believes that humans are separate from nature and that science is the way to control it. But whereas environmentalists like Rachel Carson believed that humans were being *punished* for falling from nature, Fukuyama believes that humans have been *rewarded* for separating from nature.

The problem for Fukuyama is that global warming throws a monkey wrench into his Hegelian Mechanism. Over the next fifty years, if we continue to burn as much coal and oil as we've been burning, the heating of the earth will cause the sea levels to rise and the Amazon to collapse and, according to scenarios commissioned by the Pentagon, will trigger a series of wars over basic resources like food and water.[23] Indeed, these things may happen even if we stop burning fossil fuels tomorrow. These wars could turn liberal democratic governments into military dictatorships and return humankind to a state of barbarism — thereby restarting history.

Fukuyama insists we cannot do anything to overcome global warming and that trying to do something would inevitably "kill the economic goose that is laying our technological golden eggs." Fukuyama's version of modernization theory is in this way inseparable from his market fundamentalism. For Fukuyama, the Mechanism is indivisible from the market, the means by which human societies most efficiently create the wealth and technology that allow us to dominate nature.

Both the Mechanism and the market depend upon reductive views of human nature. The Mechanism depends on the idea that the human drive to dominate nature is an essential aspect of our nature, and the market depends on the idea that the rational pursuit of our self-interest is essential. To constrain or otherwise interfere with the market in order to deal with global warming would be to interfere with the Mechanism — and that would be ecologically futile and economically destructive. The market and the Mechanism will punish those who attempt to interfere with them, Fukuyama warns.

But in the end, Fukuyama's faith in the market and the Mechanism is provisional. In his 2002 *Our Posthuman Future*, Fukuyama embraces "the political control of biotechnology." He does so because with biotechnology, as with global warming, Fukuyama's theory of *inevitable* human progress arrives at a contradiction: if the Mechanism is the expression of humankind's unquenchable drive to dominate nature, then it would only be a matter of time before we sought to control, dominate, and change our own essential nature. And if the market is the expression of our essential drive to rationally pursue our *self*-interest, then many of us would ultimately find it in our interest to change our *selves*.

Fukuyama can ignore global warming because changing and controlling nature is the whole point of the Mechanism. Biotechnology, on the other hand, cannot be ignored because it threatens to change *human* nature — and thus undermine Fukuyama's theory of History.

But both biotechnology and global warming destroy the distinctions between humans and nature on which Fukuyama's neoconservatism and environmentalism depend. Global warming is changing nonhuman natures and may end up changing us with it. It will force human societies to adapt in all sorts of ways, not the least of which could be bioengineering ourselves and our environments to survive and thrive on an increasingly hot and potentially less hospitable planet.

Ironically, Fukuyama and environmentalists like Rachel Carson ultimately share the same essentialist view that humans are driven to dominate nature, and global warming and biotechnology create profound contradictions for both. Confronted with global warming, Fukuyama throws up his hands and suggests that it offers humanity only a Hobson's choice: constrain it and bring an end to progress, or ignore it and bring an end to progress. Confronted by the logical endpoint of his own theory, that the final frontier in controlling nature would be controlling our own natures, Fukuyama calls for political constraints upon the market and the Mechanism.

Environmentalists argue that we need to change our essentially destructive natures and thus protect nature from future human violations, but then they insist that to use biotechnology to change our natures would be a violation of nature.

Both philosophies ultimately run aground upon their reductive views of both human nature, nature, and the absolutist moralities that they derive from those reductions. If essentialism is the philosophy of limits, then modern environmentalism and Fukuyama's neoconservatism are that philosophy's apotheoses.

5.

Where the fall of communism inspired one of America's most important neoconservative political philosophers to write a sweeping history of human civilization, it inspired one of America's leading liberal critics to write a simultaneously scolding and wistful microhistory of sixties-era utopianism. Paul Berman opens his *A Tale of Two Utopias* with a description of the grandiosity of youthful baby boomers:

> In the years around 1968, a utopian exhilaration swept across the student universe and across several adult universes as well, and almost everyone in my own circle of friends and classmates was caught up in it. The exhilaration was partly a fury against some well-known social injustices, and against some injustices that had always remained hidden. Partly it was a belief, hard to remember today (except in a cartoon version), that a superior new society was already coming into existence. And it was the belief that we ourselves — the teenage revolutionaries, freaks, hippies, and students, together with our friends and leaders who were five or ten years older and our allies around the world — stood at the heart of a new society.[24]

Berman understood that the longing of the young revolutionaries of the late sixties was postmaterialist, even if he never used that word specifically, and that the paroxysms of violence, the sectarianism, and the utopianism of the late sixties were also the birth of a new kind of individualism — and narcissism.

It was also, in Berman's words, "a moral crisis" — the feeling of which was captured by the 1962 Port Huron Statement, the founding document of the new left's Students for a Democratic Society (SDS), which declared, "We are people of this generation bred in at least modest comfort, housed now in universities, looking uncomfortably to the world we inherit." The statement's authors managed to sound altruistic while placing themselves at the center of everything unfolding around them.

Young liberal baby boomers imagined that with the "revolutions" of 1968, they had set in motion sweeping social and political change. But the generation of 1968 ultimately opted for continuity. They did not create a new paradigm but rather relied on the old rationalist-materialist one. The altruism, idealism, and utopianism of the new left was born of a combination of the postwar economic boom, the bipartisan liberal consensus, and the fading revolutionary dreams of their parents. The new left's Marxism was old, not new. Its critiques of corporate power and imperialism were old, not new. Its utopias were old, not new. And even its environmentalism — begun long before Earth Day — was old, not new, created by Cold War liberals and scientists, not baby boomers.

Berman began *A Tale of Two Utopias* with a tale of youthful, Dionysian excess that, though intoxicated with a peculiar authoritarianism, was still hopeful. He ended the book in a far darker mood, quoting from the French philosopher André Glucksmann's 1991 attack upon utopian politics, *The 11th Commandment,* which proposed that the new commandment be "Know thyself as capable of being a monster."[25]

Utopia originally meant "no place" but over time came to signify a real place in a necessary future. Left-wing utopians imagined that their visions of a perfectly harmonious future would be the end of history — a different end than the one described by Fukuyama, but an *end* nonetheless. Whenever people convince themselves that they know what the end of history will look like, they tend to become rather dogmatic about how to achieve it.

The problem with Berman and many other chastened liberals from the "generation of '68" is that they threw the aspirational baby out with the utopian bath water. In attacking a politics that imagines it not only possible but also desirable to remake society and human nature, Berman encourages liberals to reject even aspirations to greatness. The aspects of liberalism that Berman demands we elevate — the appreciation of ambiguity, tolerance, and compassion — are vitally important, but embracing them does not require that we embrace a small, bland, and timid humanism. Indeed, doing so in the face of the monumental challenges ahead would be irresponsible.

There can be no project of international solidarity and compassion that does not also aspire to human greatness. The new politics should have no utopia, no place, and no end. A politics of greatness demands that we aspire not to an end of history but rather to beginning new ones.

6.

In November 2006, disgusted by the bloodbath in Iraq, corruption on Capitol Hill, and incompetent Bush administration rule, Americans voted to oust Republicans from Congress. Ecstatic Democratic activists in Washington held press conferences and e-mailed their members claiming that the election was a mandate for the Democratic agenda. It was not. Rather, it was a repeal of the mandate Americans had given to Republicans in 2002 and 2004.

As voters went to the polls, few of them, Democrats included, knew what the Democratic agenda was. They weren't voting Democrats in as much as they were voting Republicans out. When Americans heard the specifics of the Democratic agenda, they mostly shrugged. Raising the minimum wage, cutting interest rates on student loans, pay-as-you-go spending — these small-bore policies aroused little passion from anyone. While raising the minimum wage is the right thing to do, fewer than one in twenty workers will be affected by it.[26] And while cutting interest rates on student loans helps some families, it is no substitute for addressing the security needs and personal aspirations of most Americans.

What's needed is a postindustrial social contract that will allow the next generation of young Americans to turn away from survival values and materialistic status competitions and toward fulfillment, self-creation, and a new ecological politics. A new politics should inspire Americans to grapple with the *existential* questions Nixon asked in his 1970 State of the Union address: What kind of a country do we want? How can we achieve it? These questions implicitly contain a question about *investment:* how will Americans invest our wealth and our labor?

In 2003 we sought to answer some of those questions by helping to create a proposal for a new Apollo project on clean energy — a version of which has since been introduced into Congress by Representative Jay Inslee. The project would invest $300 billion over ten years in the fastest-growing markets in the world: energy. Initial econometric analyses show that a portfolio of investments in wind, solar, biofuels, carbon sequestration, mass transit, hydrogen, and other energy sources would attract an additional $200 billion in private capital and create roughly three million new jobs. These investments would position the United States to take the lead in several fast-growing clean-energy markets that are currently dominated by Europe and Japan. As a result, the investments also have the potential to dramatically reduce the trade deficit; just as important, the initiative would more than pay for itself by capturing the revenues from the energy savings accrued from investments in efficiency.

But we were hardly alone in crafting an investment-centered approach to energy. New energy initiatives come from various quarters on the right and left every few months. *New York Times* columnist

Thomas Friedman seems to write a bristling column calling for energy independence every other week. In 2002, Aaron Sorkin wrote a gripping season of NBC's *The West Wing* about a fictional American president winning reelection on an Apollo-type clean-energy investment platform.

In a set of columns written in 2006, the economist Robert Samuelson argued that the solution is not emissions reductions but rather investments in new technologies.

> Unless we develop cost-effective technologies that break the link between carbon-dioxide emissions and energy use, we can't do much. Anyone serious about global warming must focus on technology — and not just assume it. Only an aggressive research and development program might find ways of breaking our dependence on fossil fuels or dealing with it.[27]

The problem with regulation-centered approaches, from Kyoto to a carbon cap-and-trade system, is not just that they are *ecologically* inadequate but also that they are economically insufficient to accelerate the transition to clean energy.

And, of course, they make for difficult politics. When environmentalists fail to raise fuel-economy standards or cap carbon, the members of Congress who voted against these initiatives pay no price during the next election, because there are so many issues that voters care more about than global warming. By contrast, the economy and national security, which are at the center of large, investment-based proposals, top the public's list of concerns. Unlike raising the minimum wage and cutting the interest for student loans, which are popular ideas but not strongly felt, support for a major energy investment is *intense*.

Part of the reason for this is that a new Apollo project tells an overarching story about America. It begins by acknowledging what America is great at: imagining, experimenting, and inventing the future. In the depths of the Great Depression, President Franklin Roosevelt became a radical experimentalist, inventing various New Deal programs to overcome hunger and joblessness. During World War II, America defeated fascism as much through our ingenuity and manufacturing muscle as through our fighting GIs. In the 1960s and 1970s, the Defense Department made a series of large and strategic investments to create the Internet, and it guaranteed the market for microchips, thereby creating the conditions for the electronics and information revolutions.

This is a story that increasing numbers of political strategists, columnists, and bloggers are telling about America, but outside of the fictional world of *The West Wing*, there are few energy independence or clean-energy proposals that put major investments at their centers. In failing to embrace investment, this story about America's future and its capacity to innovate rings false. Neither Democratic leaders in Congress nor Democratic presidential candidates can convincingly speak to American greatness as long as they refuse to put their money where their mouths are.

7.

Why do Democrats continue to put policies like minimum-wage increases and carbon caps rather than muscular investments at the top of their agendas? There are three reasons. The first is that small, incremental policies such as modest increases in the minimum wage, caps on carbon emissions, and extending health care to children (but not their parents) are the top priorities of the liberal interest groups that define the Democratic agenda. For these groups, bold, innovative, aspirational proposals to dramatically transform the economic and political landscape are seen as distractions from their attempts to advance their small, incremental policies.

Apollo, in its very conception, rejects the single-issue, incremental mindset that is virtually hard-wired into the liberal special-interest-group approach to politics and institution building. And while a pas-

sel of unions, environmentalists, and businesses have officially endorsed Apollo, none have championed it. A few have even worked behind the scenes to ensure that it never comes up for a vote. "We've
been positive publicly about Apollo," one of the environmental community's chief lobbyists told us, "but not positive policy-wise because
it doesn't have binding limits, either on CAFE or carbon."[28] Labor
unions have also endorsed Apollo but have done vanishingly little to
push for it, largely because they are too busy trying to protect jobs in
the dying old industrial economy rather than create new ones in the
postindustrial one.

The second reason for the dominant liberal politics of timidity
and the aversion to investment is that many Democrats have come to
believe, erroneously, that it was President Clinton's 1993 balanced
budget that caused the economic boom of the 1990s. The consensus
of economists, however, is that the balanced budget of 1993 had little
or nothing to do with the high economic growth of the 1990s. Rather,
it was due to a surge in productivity from the expanded use of personal computers and the Internet, supply-chain efficiencies, the globalization of production, a boom in investment, and other factors.[29]
Yet leading Democrats continue to believe that deficit reduction was
the key to the economic expansion of the 1990s, because without this
revisionism, the Clinton presidency would have, with the exception
of welfare reform, no significant domestic policy or political legacy to
speak of.

Putting People First, the domestic economic-stimulus agenda
upon which Clinton was elected in 1992, was abandoned in the early
days of his presidency. In 1994, the administration's complicated health
care reform proposal failed in Congress, and the effort ended in political catastrophe. Republicans took back both houses of Congress that
November. In 1997, the Senate rejected the Kyoto treaty, 95–0. President Clinton did not win an outright majority of the popular vote in
either 1992 or 1996. And in 2000, Clinton's vice president, Al Gore,
lost the presidential election at the height of one of the greatest eco-

nomic expansions in American history. The political mythology constructed by Clinton-era Democrats to excuse these serial political failures resulted in the valorization of timidity and the exaggeration of small political gestures.

Democrats, historically the party of social investment, are today the party of fiscal parsimony and deficit reduction. This is widely seen as a strength, despite scant evidence that it has ever been a particularly powerful political asset for Democrats. Deficit reduction served Republicans in the late 1980s and early 1990s because it was code for cutting taxes and social programs. It allowed Republicans to attack government spending without attacking specific social programs that were popular among voters. And Republicans quickly abandoned the discourse of deficit reduction once they were in control and it no longer served their purposes.

By contrast, it is not clear how the discourse of deficit reduction has served either Democratic electoral prospects or progressive social ends. Yet Democrats, even as they seized control of both houses of Congress in 2006, still hew to Clinton-era deficit dogma.

The third reason for continuing opposition to an investment-centered agenda is the dead weight of the pollution paradigm and the politics of limits. One can scarcely imagine a CEO who looks only at costs and debts and not at revenues and investments, and yet this is the mental model that several hundred members of Congress take to work every day. Market fundamentalists say that it is not the role of government to pick winners and losers in the economy. But through everything from tax policy to social policy, the government is always picking winners and losers in the economy. Nowhere is that more true than in the energy sector, where the government has for years been subsidizing coal, oil, and gas companies and virtually ignoring clean-energy companies. The problem is not that the essential market is being distorted by old energy subsidies and improper accounting, but rather that a market created to serve the old energy economy can no longer serve our present energy and ecological needs.

While "internalizing externalities," whether through carbon taxes or pollution limits, is a part of the solution, without major new investments, it alone amounts to tearing down the old energy economy before building the new one.

8.

On January 27, 2005, standing before the Davos World Economic Forum, Prime Minister Tony Blair put at the center of his vision of global interdependence two issues that had never before played such a prominent role in any geopolitical agenda: extreme poverty and global warming. "Why do it?" Blair asked. "Not just because they matter. But because on both there are differences that need to be reconciled; and if they could be reconciled or at least moved forward, it would make a huge difference to the prospects of international unity; as well as to people's lives and our future survival."

For environmentalists, Blair's Davos speech was a breakthrough. In directly challenging the Bush administration and pushing climate to the top of the G8's agenda, Blair had injected much needed political drama into what had been the sleeping pill of environmental issues. Would Blair's crusade be the tipping point environmentalists had long dreamed of?

It would not. By the time Blair hosted the July G8 summit in Gleneagles, Scotland, the Bush administration had severely diluted the joint statement on global warming. The White House removed language about rising seas and melting glaciers and other references to global warming as "an urgent threat" that required "immediate action" and inserted language that emphasized "uncertainties."[30] The prime minister tried to put his best face on the debacle by insisting that the statement's concession that *something* should be done was a sign of progress. But a White House official correctly pointed out that it was a trivial concession. Indeed, the communiqué, which stated that emissions are only "associated with" global warming, was actually weaker than the statement signed by the president's father thirteen years prior. It was then that President Bush Sr. flatly stated that emissions "result in" global warming.[31]

Though disappointed, environmentalists pointed to the commitment from all the G8 leaders except President Bush to negotiate a successor to Kyoto in Montreal in November 2005. There would be yet another opportunity, they thought, to demonstrate America's isolation on the global warming issue. But just three weeks after the Gleneagles debacle, the White House outmaneuvered Blair once again. At the Association of Southeast Asian Nations, the Bush team unveiled the Asia-Pacific Partnership on Clean Development and Climate. The partnership of China, India, Japan, South Korea, and Australia does not require mandatory greenhouse gas emissions and invests a minuscule amount of new money into new, joint technology efforts. Even so, the announcement further undermined Blair and positioned Bush as a champion of economic development in the developing world. A senior Downing Street official correctly described the secretly negotiated agreement and surprise announcement as a "slap in the face."[32]

9.

What went wrong with Tony Blair's crusade on global warming? To answer this question, we must take a closer look at the speech he gave at Davos.

"There are facts that are accepted," Blair began.

Ever since Arrhenius first predicted global warming in 1896, it has been fiercely debated. I am not a scientific expert. I only see that the balance of evidence has shifted one way. Some argue this warming is part of a natural cycle such as, by contrast, the mini ice age in the Middle Ages. But glaciers are now in retreat that have not retreated since the last Ice Age, 12,000 years ago. The impact of climate change predicted by modelers is uncannily coming to pass, not least in the European summer of 2003.

Instead of inspiring his audience around a vision of global interdependence — where economic prosperity, ecological restoration, and crosscultural understanding are woven together — Blair offered a dry, textbook description of global warming science and a defensive insistence that Kyoto wouldn't harm the economy. "However," Blair said, "behind the dispute over science is another concern. Political leaders worry they are being asked to take unacceptable falls in economic growth and living standards to tackle climate change."

It is tempting to chalk up the speech's shortcomings to uncreative speechwriting. But Blair's oratorical skills are fine — witness his powerful speeches on terrorism in the wake of September 11. Blair lacked neither the proper emotional commitment nor good speechwriters. The problem was that he viewed global warming as an environmental problem — and Kyoto as the policy solution.

If Tony Blair were the only politician to have suffered a major political setback after crusading on global warming, the episode might not be worth paying attention to. But Blair was far from the first politician and is unlikely to be the last to suffer a fall after hitching himself to the politics of limits.

10.

Almost sixty years before Blair gave his speech in Davos, Winston Churchill stood before an audience in Zurich to give his famous "United States of Europe" speech, which outlined his vision for the European Union. To fully appreciate how much more compelling Churchill's vision of a united, peaceful, and prosperous Europe is in comparison to Blair's vision for regulating greenhouse gases, it is worth quoting Churchill at length:

> We must all turn our backs upon the horrors of the past. We must look to the future . . . I am going to say something that will

astonish you. The first step in the recreation of the European Family must be a partnership between France and Germany. In this way only can France recover the moral and cultural leadership of Europe. There can be no revival of Europe without a spiritually great France and a spiritually great Germany . . .

Time may be short. At present there is a breathing space. The cannons have ceased firing. The fighting has stopped; but the dangers have not stopped. If we are to form the United States of Europe, or whatever name it may take, we must begin now . . . In all this urgent work, France and Germany must take the lead together. Great Britain, the British Commonwealth of Nations, mighty America and I trust Soviet Russia — for then indeed all would be well — must be the friends and sponsors of the new Europe and must champion its right to live and shine.

Therefore I say to you: let Europe arise!

In the sixty years after Churchill's speech, the European Union emerged as the most powerful economic force ever organized by humankind. Churchill proposed a simple first step that, magnanimously and strategically, put France and Germany — not Britain — at the center of the new union. It was a speech that helped inspire the United States to invest more than $12 billion between 1947 and 1951, the equivalent of an astonishing $500 billion in today's dollars, into rebuilding Europe and Japan.

Laboring under a common vision for Europe's future, French and German leaders spent the next several years looking for a joint economic venture that could be the basis for peace and prosperity. In 1950, France, Britain, and the United States agreed to allow Germany to start producing steel once again, with France and Belgium providing the coal, under the condition that none of the steel be used for military purposes. The energy and steel operation was wildly successful, inspiring participating nations to accelerate cooperation and integration.[33]

When historians look back on that historic joint venture they emphasize the crucial role played by the mass manufacture of steel, which was vital in rebuilding Europe. But the most salient lesson for today's leaders is the importance of common investments in new energy sources, which have historically been crucial drivers of economic development.

It is impossible to imagine Churchill ever giving a speech like the one Blair gave in Davos. Where Churchill tells a story about Europe's transition from a violently nationalist past to a prosperous and interdependent future, Blair describes the dry science of climate change and responds defensively to arguments made by market fundamentalists against action. Where Churchill declines to get into the weeds of policy development, Blair offers a laundry list of technical policy details. And where Churchill encourages us to dream about a better future, Blair speaks of scientific facts and economic fears. In focusing the world's attention on *limiting greenhouse gases* rather than on *unleashing economic potential,* Blair ended up sounding more like an environmentalist than like Winston Churchill.

11.

Perhaps the most astonishing aspect of Blair's crusade was his repeated failure to connect his clean-energy agenda to his own agenda to end extreme poverty. Instead of proposing a joint new energy partnership for economic development, Blair focused his attention entirely on the threat that developing countries like China and India will produce more carbon emissions than Europe. In failing to connect the global warming crisis to economic opportunity, Blair helped drive developing nations into the welcoming arms of the Bush administration.

Thankfully for history, Churchill didn't see the postwar rubble of Europe as a laundry list of problems to be solved by technical solutions. In the story told by Churchill, the interdependent "European Family" was the hero that would valiantly overcome "nationalistic quarrels" and put an end to the "Teutonic nations" of the past. Churchill brilliantly cast nationalism as the problem and economic integration as the solution. By contrast, Blair cast human power as the problem and science as the solution.

Davos could have been the beginning of Blair's Churchillian transformation. Instead, it was the beginning of his political end. In 2005, the prime minister became the latest casualty of the politics of limits. Instead of maximizing Britain's influence, Blair minimized it. And instead of banishing the perception that he was Bush's poodle, Blair reinforced it.

12.

What might have happened had Churchill given his speech in 1946 à la Tony Blair in 2005?

> There are facts that are accepted. The five most violent years on record have occurred in the last seven years; and ten in the last fourteen. It is over eighteen years since the world recorded a "nonviolent" month. Average monthly incomes have decreased since the 1930s.
>
> Ever since Adam Smith wrote about the economy, it has been fiercely debated. I am not an expert on the economy. I only see that the balance of evidence has shifted one way. Some argue that economic depressions are part of a natural cycle, such as the last great depression. But people are poorer now than they were then. The impact of the war on the economy predicted by the economists is uncannily coming to pass, not least in the European summer of 1944.

The absurdity of giving such a speech is hardly worth mentioning but for the sad fact that this is precisely the kind of speech that environmental leaders and their political allies give routinely. Had Winston Churchill given such a speech, Europe might be no closer to unity today than the world is to a clean-energy economy.

Now consider how things might have turned out had Blair offered a Churchillian speech in, say, Columbus, Ohio.

> I am going to say something that will astonish you. The first step in defeating the forces of terror and extremism must be a new energy partnership between the West and the East. In this way only can we recover our common moral and cultural leadership. We cannot build prosperous twenty-first-century economies in the U.S. and Europe without a spiritually and economically great Middle East and Asia.
>
> Time may be short. At present there is a breathing space. We can create millions of good new jobs by together investing billions of dollars of public and private capital to invent new, better sources of energy. America is in many ways the heart of global innovation.
>
> In all this urgent work, Britain and the United States must take the lead together. I trust that the nations of Europe, Asia, and the Middle East will be our fellow investors and friends in this shared dream.

Would such a speech have turned around George W. Bush? Almost certainly not. But had Blair contested the global warming debate with Churchillian vision rather than threatening apocalypse, counseling sacrifice, and offering policy laundry lists, he would have, at the very least, exerted far more political pressure on Bush than he did with his repeated admonitions that America eat its Kyoto peas.

Rebuffed, Blair backpedaled a few months later. He said he hoped that the nations of the world would "not negotiate international treaties," as though only national efforts on global warming made sense. What Blair and other environmentalists still do not understand is that Kyoto failed in the United States not so much because it was an international treaty but more because it imagined a future entirely in the negative.

Bush's underinvestment in his own clean development initiative was Blair's opportunity to clear a new path for action on global warming. The prime minister should have called Bush's bluff — and upped the ante — by embracing Bush's Asia-Pacific Partnership on Clean Development and proposing that Britain, the United States, and the other G8 countries fund it to the tune of $30 billion per year, creating an Opportunity Fund for Clean Development. Such a move would have helped everyone — including antipoverty activists — see the fund not simply as an environmental solution but also as a central part of the effort to make poverty history.

13.

As surely as postwar prosperity gave birth to environmentalism, global warming is transforming it. Any successful effort to stabilize the climate will destroy the distinctions between environmental protection, economic development, and global equity. China, India, Brazil, and the rest of the developing world will not agree to any international approach that constrains the economic aspirations of their people — nor should they. The average Chinese consumes 15 percent of the energy of the average American. It would be immoral to attempt to lock the developing world into energy penury.

A successful effort to stabilize the climate will thus almost certainly result in a rough equalization of per capita carbon emissions globally. Given the close connection between energy use, emissions, and living standards, the implications of this are momentous: to equalize global carbon emissions is, in the end, to equalize global living standards.

And so we arrive at a place similar to the one from where we started: the intersection of prosperity and ecological concern. But whereas postwar prosperity in the developed world created the conditions for ecological concern, ecological concern today must create the conditions for prosperity in the developing world. What were once preconditions for environmentalism must become the explicit objectives of the new politics. For environmentalism to realize its potential to transform the energy economy, it must evolve into something new. As it does, it will also transform American political life.

The new vision of prosperity will not be the vision of economic growth held by those who worship at the altar of the market. It will define wealth not in gross economic terms but as overall well-being. Wealth will be defined as that which provides us with the freedom to become unique individuals. It will embrace our power to create new markets. And it will turn the environmental movement's conditional support for economic development on its head: developing economies will be sustainable precisely to the extent that we invest in their development.

And just as dealing with global warming will become inseparable from creating global prosperity, preparing for global warming will become inseparable from preparing for the natural disasters that disproportionately plague the vulnerable global poor. Many environmentalists recognize the need for global investment and preparation but have not yet created an agenda that takes these challenges seriously. As they do, postenvironmentalism will be born.

Despite the claims of some of its adherents, environmentalism has never been above politics, and the same will be true of postenvironmentalism. Indeed, it will be deeply contested. Climate change and the political response to it is already defining a new fault line. On one side of that line will be a global NIMBYism that sees the planet as too fragile to support the hopes and dreams of seven billion humans. It will seek to establish and enforce the equivalent of an international caste system in which the poor of the developing world are consigned to energy poverty in perpetuity. This politics of limits will be anti-immigration, antiglobalization, and anti-growth. It will be zero-sum, fiscally conservative, and deficit-oriented. It will combine Malthusian environmentalism with Hobbesian conservatism.

On the other side will be those who believe that there is room enough for all of us to live secure and free lives. It will be pro-growth, progressive, and internationalist. It will drive global development by creating new markets. It will see in institutions like the WTO, the World Bank, and the International Monetary Fund not a corporate conspiracy to keep people poor and destroy the environment, but an opportunity to drive a kind of development that is both sustainable and equitable. It will embrace technology without being technocratic. It will seek adaptation proactively, not fatalistically. It will establish social and economic security as preconditions for ecological action. It will be large and transformative, but not millenarian.

These are some of the political choices that all of us will face in the coming years, for we will not put the world back the way it was, nor will we renounce our desire to control nature. We have risen, not fallen. In the words of one founding father of environmentalism, who long ago broke from the politics of limits, "We are as gods and might as well get good at it."[34]

14.

In the Greek legend, Daedalus constructs two magnificent pairs of wings out of feathers and wax for himself and his son, Icarus. As father and son fly through the air together, a plowman and a shepherd below mistake them for gods. Daedalus warns his son not to fly too close to the sun, but Icarus is so in love with his new powers that he flies higher and higher until the sun melts the wax and Icarus plunges to his death in the sea below.

The myth of Icarus has long been told to discourage audacity, shame pride, and promote timidity — especially among the young. Fall narratives are parables of limits. Whether they are Greek, Christian, or Rousseauian, these cautionary tales aim to punish transgression, inhibit innovation, and preclude possibility. Those who fear change always declare challenges to their authority to be *hubris*. They exaggerate the falling and mask the overcoming.

But there are far more people who have abandoned their dreams fearing disappointment than who have taken flight and fallen to the sea. Almost nobody believes that making the transition to a clean-energy economy and saving the Amazon are bad ideas. The problem is not that people don't see the nightmare, but rather that they do not allow themselves to dream.

The logic of dreams is expansive, ecological, and emotional. There's no distinguishing the interpretation of our dream from the dream itself. The beings in our dreams are manifestations of our feelings, selves, and ideas. Some say we are all parts of the dream; others say we are whatever we say we are. In our dreams we aren't bound by the laws of nature — not even by gravity. Good becomes evil and evil good in the blink of an eye. We fall and then we fly. We transmogrify. In a single dream you might begin as your mother, become yourself, and end as your son.

The crises we face demand not that we *wake up to reality* but rather that we *dream differently.* There is a very special kind of dreaming where one is both dreaming and reasoning at the same time. It is called a lucid dream. The lucid dreamer is in control — at least partly. This awareness gives him the power to do whatever he imagines. He can fly to the sun without fear of falling — and confront the shadow chasing him.

For much of human history we have told ourselves stories aimed at constraining our potential out of fear of offending higher political authorities. But today, there is no political authority higher than humankind itself. Whether we like it or not, humans have become the meaning of the earth.

We should see in hubris not solely what is negative and destructive but also what is positive and creative: the aspiration to imagine new realities, create new values, and reach new heights of human possibility. And we should see in humility not timidity but *gratitude* — a gratitude for the achievements of our ancestors, the emergence of our species, and the gift of our existence.

If global warming and the destruction of the Amazon require a kind of greatness — even hubris — humankind has never before seen, they also require transcending outmoded ways of being in the world. We must see our multiple human and nonhuman natures as neither inherently good nor inherently bad. We must nurture those natures that help us to overcome our lower needs and values.

Humans didn't evolve to deal with great challenges like global warming, it is true. But neither did we evolve to overcome deprivation, create beauty, or achieve happiness. In overcoming oppression and deprivation — predators, hunger, disease — we have given birth to a new world. It is a world at once beautiful and terrible. And this world, too, we shall overcome.

In Gratitude

This book would not have been written had it not been for the encouragement of our family and friends. Peter Teague has for many years asked tough, imaginative questions, and demanded ever more audacious answers. Irene Hughes has helped us to see ourselves more clearly and engage our allies and our opponents more constructively. Adam Werbach's courage and enthusiasm constantly inspire us.

We are grateful to Lance Lindblom, Sara Kay, Adam Cummings, and the board of the Nathan Cummings Foundation for their steadfast support, and to Bracken Hendricks for working with us to develop Apollo, Health Care for Hybrids, and Global Warming Preparedness.

Ross Gelbspan, Jon Isham, and Bill McKibben, vital writers and teachers on global warming, were courageous in their defense of Cape Wind, and have always encouraged us even when we disagreed.

Cara Pike, Georgia McIntosh, Trip Van Noppen, Buck Parker, and the staff and board of Earthjustice were always willing to ask hard questions and seek unconventional answers to ecological crises. Celinda Lake, David Mermin, and the whole Lake Research team were terrific partners in our health care research collaboration.

Pamela Morgan, Kenton DeKirby, John Whaley, and Jenn Bernstein were invaluable colleagues in our joint research into changing social values, global warming preparedness, health care, and attitudes toward environmentalists. Erin Malec and Aden Van Noppen were scrupulous researchers and readers. Jeff Navin and Kelly Young encouraged our visions and shaped our thinking. And we benefited enormously from seeing America through the eyes of our Canadian social values research colleagues at Environics: Michael Adams, Barry Watson, David Jamieson, Derek Rustow, and Jan Kestle.

In addition to being one of the most brilliant and fearless journalists ever to have reported on the Amazon, Leonardo Coutinho was exceptionally generous and patient in his explanations of the situation unfolding inside and outside the forest.

Bill Chaloupka is a joyful scientist, critic, and friend — we are grateful to him for never ceasing to challenge our thinking. Jim Proctor offered important insights into the relationship between science and religion. We appreciate both men for having invited the wider academic community to discuss the future of postenvironmentalism.

We thank Anton Mueller for his wise counsel, cogent criticism, and invaluable contributions, and the entire Houghton Mifflin team. Thanks to Rafe Sagalyn for helping us navigate the looking-glass world of publishing with aplomb.

Several people reviewed versions of the manuscript and offered excellent counsel: Michael Adams, Kyle Danish, Jonathan Diskin, Christopher Foreman, Mott Greene, Kathleen Marie Higgins, David Hoy, Frank Laird, Geoff Lomax, John McWhorter, Pamela Morgan, Perry Morse, Bill Nordhaus, Bob Nordhaus, Hannah Nordhaus, Jean Nordhaus, Dara O'Rourke, Roger Pielke Jr., Karin Rosman, Daniel Sarewitz, Bob Shellenberger, Daniel Silverman, Les Thiele, John Whaley, and Robb Willer.

Our thanks go to those environmentalists who debated with us, especially Carl Pope, the first environmental leader to publicly criticize our essay and seriously argue with its contents.

Ted

My parents, Bob and Jean, taught me to think and write, and always supported my unconventional career choices. My sister Hannah always asks hard questions and helped us rethink how to organize the book. My wife Sara's love emboldens me to dream. Without you all, this book would never have been possible.

Barry Rosenberg, Danny Cantor, Doug Phelps, Barry Nelson, Doug Linney, Deborah Raphael, and Alex Evans have been friends, teachers, and mentors.

Geoff Lomax, Daniel Silverman, Tom Riley, Joe Richman, Matt Baker, Scott Smith, Aana Lauckhart, Shelley Fidler, and Jackie Wallace are but a few in the community of friends who have engaged and challenged the ideas that became this book, and have stood by me and encouraged me through good times and bad.

My grandfather Robert J. Nordhaus was a man of ideas and action. He read this book in manuscript form shortly before his passing. It was the last book he read, and his passion for life and fortitude over ninety-seven remarkable years will always inspire.

Michael

Many Brazilians were generous in sharing their country with me while offering their friendship: Marcia Meirelles, Fernanda Pessoa, Maria Luisa Mendonça, Marina Thereza Campos, Erotilde Honório Silva e Ivo Roza da Silva, Maria Gorete Sousa, and the perceptive, courageous Eduardo Borges de Oliveira.

Several teachers had a profound influence on my spirit and my thinking: Jon Diskin, Caroline Higgins, JoAnn Martin, Howard Richards, Kathy Taylor, and especially the indomitable Dr. Ken Miller, an audacious visionary and true friend. I am grateful to Erik Curren, Marc Lavine, Tony Newman, Matt Rosen, Richard and Livia Rosman, and Kim Shellenberger for being wise, caring, and resolute friends. And I am thankful every day for the gifts of my beloved Karin and our children, Joaquin and Kestrel.

This book is dedicated to my parents: Nancy and Don O'Brian, Bob Shellenberger and Judith Green. They gave me life and stirred in me the capacity to take hold of it.

Notes

Introduction

1. *Guardian*, "I Have a Dream," August 21, 2003.
2. Interview with Sherry Dupree, the archivist for the Gospel Hall of Fame and Museum in Detroit, National Public Radio, *The Tavis Smiley Show*, August 28, 2003. Full song lyrics at www. negro spirituals.com/news-song/i_ve_been_buked_and_i-ve_been_scorned .htm.
3. Full transcript of Martin Luther King Jr.'s "I Have a Dream" speech, delivered August 28, 1963, can be found at www.americanrhetoric.com/ speeches/mlkihaveadream.htm.
4. *Tavis Smiley Show*, August 28, 2003. Also see Taylor Branch, *Parting the Waters* (New York: Touchstone, 1998), 882.
5. *Guardian*, "I Have a Dream."
6. Branch, *Parting the Waters*, 883.
7. Jerome Armstrong and Markos Moulitsas Zúniga, two prominent progressive bloggers, recently observed, "The Democratic Party stands for everything, yet stands for nothing . . . Whatever successes they had were in the 1960s and 1970s. For too long now these single-issue groups have been fighting not to advance their causes, but to stop the backslide." In *Crashing the Gate* (White River Junction, VT: Chelsea Green, 2006), 37, 38.
8. A. H. Maslow, "A Theory of Human Motivation," *Psychological Review* 50 (1943): 370–96.
9. The two of us are research, policy, and strategy consultants to environ-

mental and progressive foundations and organizations, through both our research and our strategy firm, American Environics, which is a joint venture with the Canadian-social values research firm Environics, and through the Breakthrough Institute, our nonpartisan think tank.

10. Emissions increased 4.1 percent between 2000 and 2004. United Nations Climate Change Secretariat, "2006 UNFCCC Greenhouse Gas Data Report Points to Rising Emission Trends," October 30, 2006.

11. Rob Collier, "A Warming World: China about to Pass U.S. as World's Top Generator of Greenhouse Gases," *San Francisco Chronicle,* May 5, 2007.

12. Measuring carbon emissions from deforestation is notoriously tricky. The 25 percent figure comes from the United Nations International Panel on Climate Change's "Climate Change 2007: The Physical Science Basis — Summary for Policymakers," February 2007, www.ipcc.ch/SPM2feb07.pdf.

13. Last year, Indonesia became, after the United States and China, the third largest emitter of greenhouse gases, mostly through burning forests and peatland in order to grow, paradoxically, cleaner-burning biofuels. Elisabeth Rosenthal, "Once a Dream Fuel, Palm Oil May Be an Eco-Nightmare," *New York Times,* January 31, 2007.

14. "The Democratic left wants it to be 1968 in perpetuity; the Democratic center wishes for 1992 to repeat itself over and over again. History, however, doesn't oblige such wishes — it rewards those who recognize new moments as they arise," Michael Tomasky aptly wrote in his "Party in Search of a Notion," *American Prospect,* May 2006, 28.

15. Alexis de Tocqueville, *Democracy in America,* trans. Gerald Bevan (New York: Penguin, 2003), 594.

1. THE BIRTH OF ENVIRONMENTALISM

1. Quoted in Jonathan H. Adler, "Fables of the Cuyahoga: Reconstructing a History of Environmental Protection," *Fordham Environmental Law Journal* 14 (2002): 91.

2. Ibid., 101–3. Adler found reports of fires in 1868, 1883, 1887, 1912, 1922, 1936, 1941, 1948, and 1952 — and many other fires may be lost to history.

3. By the time newspaper photographers and local TV news crews arrived, the Cuyahoga River fire of 1969 was already out. "The photo in

the *Cleveland Plain Dealer* showed a fireboat spraying down a railroad trestle after the fire was under control. The *Cleveland Press* could only run a photo of the railroad ties warped by the heat of the flames." Ibid., 97–98.

4. Ibid., 101, 105.
5. Ibid., 99.
6. Our emphasis. François Leydet, *The Last Redwoods and the Parkland of Redwood Creek* (San Francisco: Sierra Club, 1963), 132.
7. For an excellent discussion of material versus postmaterial values, see Ronald Inglehart, *Modernization and Postmodernization* (Princeton, NJ: Princeton University Press, 1997), and Ronald Inglehart and Christian Welzel, *Modernization, Cultural Change, and Democracy* (Cambridge, England: Cambridge University Press, 2005).

Ronald Inglehart and his colleagues have been conducting social-values research in eighty-one societies since 1981 as part of the University of Michigan's World Values Survey. The data represent 85 percent of the world's population.

Inglehart finds that social values and economic development progress hand in hand in predictable ways. He rejects the notion that either economic development or social values drive modernization and argues that the two are inextricably linked.

"Early versions of modernization theory were too simple . . . Sociocultural change is not linear. Industrialization brings rationalization, secularization, and bureaucratization, but the rise of the knowledge society brings another set of changes that move in a new direction, placing increasing emphasis on individual autonomy, self-expression, and free choice. Emerging self-expression values transform modernization into a process of human development, giving rise to a new type of humanistic society that is increasingly people-centered." Inglehart and Welzel, *Modernization, Cultural Change*, 1.

8. A. H. Maslow, "A Theory of Human Motivation," *Psychological Review* 50 (1943): 370–96. Also see A. H. Maslow, *The Farther Reaches of Human Nature* (New York: Viking, 1971).
9. Some point to countries like China, where people seem to put their belonging needs before their physiological or safety needs. Others argue that the lines between different needs are arbitrary. Given that we gain much of our self-esteem in relationships with others, what's the

difference between an esteem need and a belonging need? And still others challenge Maslow's contention that higher needs won't emerge until lower needs are satisfied. Isn't it possible to, for example, seek self-actualization even if you live in a repressive dictatorship?

These are important challenges, but not ones that undermine Maslow's very basic theory that (a) we can distinguish between lower needs and higher needs, and (b) higher needs do not emerge strongly, in either individuals or societies, until lower needs have been adequately met. You do find self-actualizing individuals living in dictatorships where they are deprived of political freedom, but you do not find nearly as many of them as you do in freer societies. And those self-actualizing individuals living in dictatorships are usually more affluent and thus often have greater freedoms than their poorer countrymen and -women.

A stronger argument can be made against the labeling of specific levels in Maslow's pyramid of values. Indeed, oftentimes belonging needs *are* esteem needs — one might get self-esteem from belonging to a particular group, for example. But Maslow was writing of a much more basic sense of self-esteem, the fundamental sense of one's personal power to realize and overcome one's needs and confront new ones.

10. Throughout this book we use the word *self-creation* instead of Maslow's term *self-actualization. Self-actualization* has the connotation that a person's potential exists as an internal essence from birth, whereas *self-creation* recognizes the contingent and creative nature of the process Maslow describes.

Today, even psychologists most skeptical of Maslow's writings around self-actualization acknowledge the importance of his work and humanistic psychology in general. This passage is from a widely used psychology textbook from 1995.

The humanistic psychologists have reminded us of many phenomena that other approaches to the study of personality have largely ignored. People do strive for more than food and sex or prestige; they read poetry, listen to music, fall in love, have occasional peak-experiences, try to actualize themselves. Whether the humanistic psychologists have really helped us to understand these phenomena better is debatable. But . . . what they have done is to insist that these phenomena are there, that they constitute a vital aspect of what makes us human, and that they must not be ignored.

Cited in Richard Lowry, foreword to *Toward a Psychology of Being,* by A.H. Maslow, (New York: John Wiley and Sons, 1999), xxxiii.

11. "The emergence of Postmaterialism," writes Inglehart, "does not reflect a reversal of polarities, but a change of priorities: Postmaterialists do not place a negative value on economic and physical security — they value it positively, like everyone else; but unlike Materialists, they give even higher priority to self-expression and quality of life." In Inglehart, *Modernization and Postmodernization,* 35.

12. Our use of the term *inner-directed* should not be confused with the way David Riesman used the same term in his 1952 sociological classic *The Lonely Crowd,* which described three types of Americans: the tradition-directed, the inner-directed, and the other-directed.

    For Riesman, the inner-directed person was fairly conservative, acting on values set in childhood, whereas the other-directed person was more-status oriented, directed toward pleasing others. These terms, which became popular in the fifties and sixties, are no longer very useful for describing different levels of postmaterial needs.

13. Elsewhere Inglehart observes, "The relatively favorable attitude of Postmaterialists toward environmental causes is not just a matter of lip service: their behavior reflects their distinctive values to an even *greater* extent than do their attitudes. Although Postmaterialists are only about twice as likely as Materialists to favor environmental protection, they are four to 10 times as likely to be active members of environmental protection groups. And Postmaterialists are four to six times as likely to vote for environmentalists parties (in countries that have them) as are Materialists." Inglehart, *Modernization and Postmodernization,* 242.

    Inglehart writes, "In most societies, the Green activists are mainly postmaterialists, and it seems unlikely that Green parties or environmentalist movements would have emerged without the intergenerational cultural changes that gave rise to a postindustrial worldview that reflects an increased awareness of ecological risks." Inglehart and Welzel, *Modernization, Cultural Change,* 39–40.

    See also ibid., 25.

14. In response to the question "If you disagreed with a candidate on just that issue, would you still consider voting for that candidate, or would you not vote for that candidate based on that issue alone?" 53 percent named gay marriage, 51 percent named abortion, 49 percent

named illegal immigration, and 44 percent named the environment as issues that would cause them to vote against a candidate. Survey for the Nicholas Institute of Environmental Policy Solutions, Peter Hart Research and Public Opinion Strategies, eight-hundred-person survey of voters, August 25–28, 2005.

15. Ibid.

16. Pew Research Center for the People and the Press, "Partisanship Drives Opinion: Little Consensus on Global Warming," public-opinion survey of 1,501 adults, June 14–19, 2006. Pew Center, "Public Views Unchanged by Unusual Weather," January 24, 2007.

17. Thirty-eight percent named the economy, 28 percent named unemployment, 11 percent health care, 10 percent the war in Iraq, 7 percent terrorism, 5 percent lack of religion or morality, 5 percent manufacturing/business/jobs going overseas. The poll was conducted among four hundred likely voters in Pennsylvania by the Evans/McDonough Company, April 22–24, 2003.

18. Environics Research Group, "Socio-Cultural Trends: 3SC," 2004, www.americanenvironics.com.

19. David Brooks, "A House Divided, and Strong," New York Times, April 5, 2005.

20. The insecurity people experience appears to be based on personality, social values, and situations. In other words, some people are more likely to feel insecure than others, while certain situations can create feelings of insecurity.

   At the level of social values, Inglehart observes, "[T]he collapse of security would lead to a gradual shift back toward materialist priorities." Inglehart and Welzel, Modernization, 35.

   In a major review of social science research since the 1950s, a group of social scientists found that fear of death, change, ambiguity, complexity, and system collapse were some of the most powerful predictors of conservatism. This confirmed other research showing that people hold particular political views to meet psychological needs. In situations where people fear events such as system collapse, they tend to become more right-wing and authoritarian. The meta-analysis was of 88 studies in 12 countries with a total of 22,818 cases studied. John T. Jost et al., "Political Conservatism as Motivated Social Cognition," Psychological Bulletin 129, no. 3 (2003): 339–75.

   The authors concluded, "Conservative ideologies, like virtually all other belief systems, are adopted in part because they satisfy various psychological needs. To say that ideological belief systems have a strong motivational basis is not to say that they are unprincipled, unwarranted, or unresponsive to reason or evidence. Although the (partial) causes of ideological beliefs may be motivational, the reasons (and rationalizations) whereby individuals justify those beliefs to themselves and others are assessed according to informational criteria." Kruglanski, 1989, 1999, 369.

The authors add, "We regard political conservatism as an ideological belief system that is significantly (but not completely) related to motivational concerns having to do with the psychological management of uncertainty and fear. Specifically, the avoidance of uncertainty (and the striving for certainty) may be particularly tied to one core dimension of conservative thought, resistance to change . . . Similarly, concerns with fear and threat may be linked to the second core dimension of conservatism, endorsement of inequality." Ibid.

21. Ronald Inglehart, Mansoor Moaddel, and Mark Tessler conducted a survey of 2,300 adults in Iraq in 2004 and found that Iraqis, in the words of David Brooks, are "a people who, buffeted by violence, have withdrawn into mere survival mode. They are suspicious of outsiders and intolerant toward weak groups, and they cling fiercely to what is familiar and traditional." David Brooks, "Closing of a Nation," *New York Times*, September 24, 2006.

22. See Michael Adams, *American Backlash* (New York: Penguin, 2005).

23. See Jacob Hacker, *The Great Risk Shift* (New York: Oxford University Press, 2006).

24. The election of a progressive during the Great Depression, Franklin Delano Roosevelt, is sometimes seen as an exception to this pattern. But we should keep in mind the extent to which Roosevelt was as authoritarian as he could be, under the law and sometimes outside it. The most famous example of this was his sending of Japanese Americans to concentration camps.

2. THE FOREST FOR THE TREES

1. Michael S. Serrill and Ian McCluskey, "Unholy Confession," *Time*, May 13, 1996; Mac Margolis, "Nightmare for Brazil," *Los Angeles Times*, September 4, 1993.

2. This despite the fact that additional pressure came in the form of a

new massacre one month later, when forty off-duty cops shrouded in black hoods machine-gunned down twenty-one random passersby in Vigario Geral, a poor neighborhood in Rio. The cops were seeking revenge for four comrades who had been gunned down the day before by drug traffickers. Tracking down the killers would have been difficult and dangerous, so the police took their rage out on ordinary innocents — retirees, porters, mechanics, a fifteen-year-old girl — who had made the fatal mistake of walking through their neighborhood at a moment when justice needed to be served. Sebastian Rotella, "Brazil Tries to Rein In Its Police," *Los Angeles Times*, August 27, 1996.

3. Tony Smith, "Child Poverty Still Plagues Latin America," Associated Press, September 16, 2001.

4. Sometimes the traumatized come back for revenge. On June 12, 2000, Sandro do Nascimento, a survivor of Candelária, held up midday traffic in Rio de Janeiro by hijacking a bus and threatening to kill everyone on board. The incident inspired an extraordinary 2002 documentary — *Bus 174* — winning international awards and critical accolades.

5. A Nexis database search revealed a single AP story about the massacre, aptly entitled "Brutal Massacre in Rio Slum Gets Little Attention Due to Rich-Poor Divide and Pope's Death," written a week later. Michael Astor, Associated Press, April 8, 2005.

6. The violence against children is just the tip of the iceberg: according to a UN study, 500,000 Brazilians were killed by guns between 1979 and 2003 — four times as many as have died in the Arab-Israeli conflict in the last fifty years. Steve Kingstone, "UN Highlights Brazil Gun Crisis," *BBC News*, June 27, 2005.

7. Juliana Resende, "Gunshot Wounds 'Endemic' in Rio de Janeiro," *San Francisco Chronicle*, November 18, 2005.

8. Larry Rohter, "New Attacks by a Heavily Armed Gang Rattle Brazil and Set Off Political Wrangling," *New York Times*, August 13, 2006.

9. *Notes from a Personal War*, documentary film, on DVD of *City of God*, 2001.

10. Ibid.

11. Harry Vanden, "Brazil's Landless Hold Their Ground," *NACLA Report on the Americas*, March 1, 2005.

12. Ibid.

13. Thomas Skidmore, *Brazil: Five Centuries of Change* (New York: Oxford University Press, 1999), 147.

14. See Phyllis Parker, *Brazil and the Quiet Intervention* (Austin: University of Texas Press, 1964, 1979), and Jan Knippers Brack, *United States Penetration of Brazil* (Philadelphia: University of Pennsylvania Press, 1977).

15. Quoted in Susanna Hecht and Alexander Cockburn, *The Fate of the Forest: Developers, Destroyers and Defenders of the Amazon* (New York: Harper Perennial, 1990), 104.

16. For discussions of the history of Amazon colonization, see Skidmore, *Brazil;* Hecht and Cockburn, *Fate of the Forest*; Marianne Schmink and Charles H. Wood, *Contested Frontiers in Amazonia* (New York: Columbia University Press, 1992); Andrew Revkin, *The Burning Season* (Washington, DC: Island Press, 1990, 2004); Aziz Ab'Sáber, "Problemas da Amazônia brasileira," *Estudos Avançados* 19, no. 53 (2005); Bertha K. Becker, "Amazônia: Desenvolvimento, Governabilidade e Soberania, Documento para o IPEA [Instituto de Pesquisa Econômica Aplicada]: O Estado da Nação," in *Brasil: o estado de uma nação,* ed. Fernando Rezende and Paulo Tafner (São Paulo: Instituto de Pesquisa Econômica Aplicada, 2005). See also Bertha K. Becker, "Geopolítica da Amazônia," *Estudos Avançados* 19, no. 53 (2005).

17. Skidmore, *Brazil,* 147–49, 178. Inflation figures are from Werner Baer, *The Brazilian Economy,* 4th ed. (Westport, CT: Praeger Publishers, 1995), 392–93.

18. Skidmore, *Brazil,* 180–81.

19. Ibid.

20. Fabio Alves and Adriana Brasileiro, "Brazil Fixed-Rate Debt to Exceed Floating in 2007," *Bloomberg News,* January 17, 2007.

21. Latin America as a whole has also paid off the equivalent of its debt in interest payments, more than $730 billion between 1982 and 1996, but it has still not reduced its debt inventory. Alicia Asper, "Latin American Debt Relief: There Is Less Than Meets the Eye," Council on Hemispheric Affairs Analysis, August 2, 2005.

22. Daniel C. Nepstad, Claudia M. Stickler, and Oriana T. Almeida, "Globalization of the Amazon Soy and Beef Industries: Opportunities for Conservation," *Conservation Biology* 20, no. 6(December 2006): 1595.

23. Censos Demográficos, IBGE, cited in Becker, "Amazônia."

24. Leonardo Coutinho, e-mail correspondence, November 3, 2005.

25. In his reporting, Coutinho notes that only 25 percent of the budget

pledged for environmental law enforcement was ever released — enough to last until August 2005, precisely the time that the burning season began. If the Amazon was divided equally among the 695 officers assigned to enforce environmental laws in the region, each officer would be responsible for an area five times larger than the megalopolis of São Paulo. And 80 percent of the fines levied by the environmental agency are never paid. Leonardo Coutinho, "As 7 Pragas da Amazônia," *Veja*, October 12, 2005, 111–12.

26. Leonardo Coutinho, telephone interview with Mark Shellenberger, June 6, 2005.

27. Numbers are from the Brazilian government. Kevin G. Hall, "Slavery Exists Out of Sight in Brazil," Knight-Ridder Newspapers, September 5, 2004.

28. Ibid.

29. For discussion of law enforcement failure in the Amazon, see Leonardo Coutinho, "Fiscal, espécie rara," *Veja*, January 28, 2004. "With about 54 million hectares protected by law in 250 federal reserves, the law enforcement branch of the Environment Ministry relies on 1,483 agents. With each one responsible for monitoring an area of 36,400 hectares, the size of Belo Horizonte, but in a region of forests, mountains, and gorges."

The same is true for the rest of the world. According to the International Tropical Timber Organization, all 95 percent of the world's remaining tropical forests exist without actual physical protection, even though two-thirds are under some sort of legal protection. Lisa Adams, "Tropical Forests Still Unprotected, Report Concludes," Associated Press, May 25, 2006. For original report, see www.itto.or.jp/live/PageDisplayHandler?pageId=217&id=1262.

30. Emphasis ours. Hecht and Cockburn, *Fate of the Forest*, 216. In the authors' own words: "Tropical exuberance honors moral perfection" (11).

31. Schmink and Wood, *Contested Frontiers*, 16.

32. Revkin, *Burning Season*, xiv.

33. Ibid., 261.

34. Osmarino Amâncio Rodrigues, "O Segundo Assassinato de Chico Mendes," *Cartas da Amazônia*, 1990. Cited in Schmink and Wood, *Contested Frontiers*, 16.

35. Coutinho interview, June 6, 2005.

36. "Roughly 40 percent of the Brazilian Amazon is under some form of protection, which is impressive," Thomas Lovejoy told the *Folha de*

São Paulo, "Americano Thomoas Lovejoy, pioneiro na pesquisa da Amazônia, diz que transformar árvores em CO_2 afeta soberqaniá," *Folha de São Paulo,* June 23, 2002.

The Nature Conservancy's David Cleary told us, "Roughly 20 to 25 percent of the Brazilian Amazon is indigenous areas, formally recognized reserves. Another 10 percent is national parks system. That takes you to 35 percent. Then there are sustainable development reserves. That adds up to 40 percent protected in some form. Sixty percent is in private hands, or theoretically in private hands. A big hunk of that is public land but could be deeded to private owners. So if 80 percent of that could be preserved, that's not bad at all. The Amazon is bigger than the continental U.S., so 20 percent is enough for development needs and enough for economic and jobs demands." David Cleary, telephone interview with M.S., April 27, 2005.

But if push comes to shove, there's no reason to believe that the Brazilian government would favor the Amazonian Indians over miners, ranchers, or farmers.

For discussion of conflicting priorities, see Francisco de Assis Costa, "Questão agraária e macropolíticas para a Amazônia," *Estudos Avançados* 19, no. 53 (2005); Leonardo Coutinho, "A floresta dá dinheiro," *Veja,* August 22, 2002; and "A Amazônia será ocupada: entrevisa com David McGrath," *Veja,* November 12, 2003.

37. At the time of this writing, private landowners had to conserve 80 percent of their forestlands and the Brazilian congress has been debating reducing the requirement to 50 percent in exchange for stronger law enforcement.

38. Anthony Anderson, telephone interview with M.S., August 17, 2005.

39. Numbers from Brazil's Instituto Nacional de Pesquisas Espaciais.

40. Coutinho, "As 7 Pragas da Amazônia," 104–5.

41. Instituto Nacional de Pesquisas Espaciais, "Deforestation Estimates for the Brazilian Amazon," São José dos Campos, Brazil, 2000. Cited in W. F. Laurance et al., "The Future of the Brazilian Amazon," *Science,* January 19, 2001, 438.

42. Based on data from the U.S. Department of Agriculture, Food and Agriculture Organization, Institute for International Trade Negotiations, and cited in Jerry Hirsch and Henry Chu, "Brazil's Rise as Farming Giant Has a Price Tag," *Los Angeles Times,* August 22, 2005.

43. W. F. Laurance, A.K.M. Albernaz, P. Fearnside, H. L. Vascon-

celos, and L. V. Ferreira, "Deforestation in Amazonia," *Science* 304, 1109–11.

Some scientists have criticized Laurance et al.'s focus on Brazil's infrastructure projects like road paving as the main cause of deforestation, pointing to the widespread deforestation in places like São Felix do Xingu, Pará, where "cattle farmers and local municipal governments build unpaved roads themselves," concluding that deforestation is "more localized and much less dependent on federal government infrastructure investments in the 1970s and 1980s." These disagreements are principally about how much, not whether, government infrastructure projects are a proximate driver of deforestation.

The debate can be found in the journal *Science* 307, 1043–44.

Some scientists argue that the focus on infrastructure misses the "interaction of cultural, demographic, economic, technological, political and institutional issues." See R. Schaeffer and R.L.V. Rodrigues, "Underlying Causes of Deforestation," *Science* 307, 1046–47.

44. Other estimates put the figure at $45 billion, though the Brazilian Embassy said the planned total was only $12 billion, of which $8 billion would go to infrastructure projects. But a second totaling of planned infrastructure investments, based on the government's three-volume, 703-page project report released in 2000, arrived at the $20 billion figure for "projects that are much more likely to affect forests." In W. F. Laurance et al., "Letters," *Science* 291.

45. Ibid.; Nepstad et al., "Globalization," 1595–1603.

46. Cited in P. M. Fearnside et al., "A Delicate Balance in Amazonia," *Science*, February 18, 2005, 1044–45.

47. Editorial, "Amazon at Risk," *New York Times*, May 31, 2005.

48. The scientists developed models that used satellite data to model an optimistic and "nonoptimistic" deforestation for the Brazilian Amazon by 2020. Under the optimistic scenario, 25 percent of the Brazilian Amazon would be deforested; under the nonoptimistic scenario, that estimate climbs to 28 percent, or 42 percent counting "seriously degraded by logging, ground fires, forest fragmentation, edge effects, overhunting, mining and other activities." This point was clarified in a letter by W. F. Laurance et al., *Science*, February 1, 2001.

Stephen Schwartzman and Robert Bonnie of Environmental Defense argued in a letter to *Science* on May 31, 2001, that an earlier study by G. Carvalho, A. C. Barros, P. Moutinho, and D. Nepstad published in *Nature* 409, no. 131 (2001), and a paper, "Avança Brasil, Os Custos Ambientais para Amazônia," published by the research institute IPAM in 2000, offered a more reliable, conservative prediction that roughly one-third of the Amazon would be deforested by 2030.

W. F. Laurance et al. responded in *Science* on May 31, 2001, that their later study's model was superior and in fact more conservative, as it was based on extrapolations from all Amazonian highways, not just the four chosen by Carvalho et al., and because it considers other infrastructure projects not included in the Carvalho study. The debate over modeling technique will undoubtedly continue indefinitely, with all authors recognizing the challenges in predicting future deforestation.

49. Laurance et al., "The Future of the Brazilian Amazon."
50. José Paolo Silveira, "Development of the Brazilian Amazon," *Science* 292.
51. Coutinho, "As 7 Pragas da Amazônia," 108.
52. Alex Bugge, "Brazilians Fear Riches May Spark Invasion," Reuters, June 5, 2005. The poll was conducted among two thousand Brazilians, April 8–13, 2005, by the polling firm IBOPE for the animal welfare group Renctas.
53. T. E. Lovejoy, "Aid Debtor Nations Ecology," *New York Times*, October 4, 1984.
54. Thomas Lovejoy, telephone interview with M.S., August 23, 2005.
55. See for instance Becker, "Amazônia": "Access to our biodiversity is free and thus favoring 'biopiracy,' indicating the urgent need for regulation of this market and the persistent use of advanced technologies."
56. Joseph A. Page, *The Brazilians* (Reading, MA: Perseus, 1995), 295; Anderson interview, August 17, 2005.
57. John Terborgh, *Requiem for Nature* (Washington, DC: Island Press, 1999), 198. "The notion that biodiversity is a global commons, belonging only to the planet earth, is not new. To date, however, governments and international bodies have failed to take the idea seriously," he adds.
58. E. O. Wilson calls Terborgh "a distinguished biologist" and claims that *Requiem* was written with "compelling documentation." Jared Diamond calls the book a "must read" and refers to Terborgh as "one of the world's greatest field biologists."

59. "What is absolute, enduring, and irreplaceable is the primordial nourishment of our psyches afforded by a quiet walk in an ancient forest or the spectacle of a thousand snow geese against a blue sky on a crisp winter day," Terborgh adds. *Requiem*, 19.

60. John Terborgh, telephone interview with M.S., April 20, 2005.

61. Terborgh, *Requiem*, 38.

62. Tyler Bridges, "Fugitive Fujimori Plans Run for Peru Presidency," *Miami Herald*, August 19, 2005.

 Bridges writes, "'In all, 42 people have been convicted of crimes involving the Fujimori government,' said special anti-corruption prosecutor Antonio Maldonado. He said the Toledo government has collected $162 million illegally stashed abroad."

63. When we pressed Professor Terborgh on his unabashed Fujimori boosterism, he said, "Fujimori changed the country in many, many ways that are almost all to the good, but he got tangled up with [intelligence chief] Montesinos."

 But Peruvian prosecutors pointed to strong evidence that Fujimori knew full well about Montesinos's death-squad activities. Javier Ciurlizza, special legal adviser to the Foreign Ministry, told the *New York Times*, "Fujimori is not just another Latin American president who is in trouble. We are in front of a mafia, the leader of a mafia."

 As of October 25, 2005, Fujimori faced eighteen corruption and four human rights charges in Peru. James Brooke, "Peru's Fugitive Ex-Leader Trying to Regain the Presidency," *New York Times*, October 25, 2005.

 Also see reports by Bridges, "Fugitive Fujimori"; Jack Change, "Peru Spy Chief's Trial Begins with a Delay," *Miami Herald*, August 21, 2005; and Hanna Hennessy, "Peru Spy Chief Faces Abuse Trial," *BBC News*, August 17, 2005.

64. Terborgh, *Requiem*, 170.

65. Ibid., 53.

66. This information per the World Resources Institute can be found at http://earthtrends.wri.org/pdf_library/country_profiles/pop_cou_604 .pdf and http://earthtrends.wri.org/pdf_library/counry_profiles/pop_cou_840.pdf.

67. Terborgh, *Requiem*, 56.

68. Ibid., 57.

69. Terborgh thus decries in no uncertain terms efforts to help the poor.

"In the United States, government-sponsored experiments in social engineering have expended hundreds of billions of dollars in funds and yet have achieved only modest progress toward the goal of reducing poverty. The social problems of the big cities remain stubbornly intractable. Attempting to carry out social engineering in a foreign culture that is undergoing rapid social and economic change within the period of a five-year assistance project can, in my view, be likened to pouring water into the Sahara Desert." Ibid., 203.

70. Ibid., 188.

71. Terborgh calls these peaceful occupations "*invasiones*" — the word used by Latin America's landlords and ruling elite — and labels them "unlawful." In fact, many Latin American governments, including Brazil's, allow for the peaceful occupation of idle land, which often leads to its purchase or expropriation so that it can be used for growing food. There's little question where Terborgh's sympathies lie: "Landless migrants to the city gather together, target a vacant plot, and, on an appointed night, rush in to erect makeshift shanties of cardboard and sheet metal before dawn. The next morning, the landowner is faced with a *fait accompli*." Ibid., 167.

72. Kim MacQueen, "Roads to Ruin," *FSU Research in Review,* Summer 1994.

73. *New York Times,* "Brazil's Leader Speaks Out," February 7, 2007.

74. Coutinho, "As 7 Pragas da Amazônia," 108.

3. Interests Within Interests

1. "In communities with two or more facilities or one of the nation's five largest landfills, the average minority percentage of the population was more than three times that of communities without facilities (38 percent versus 12 percent)." United Church of Christ, Commission on Racial Justice, "Toxic Wastes and Race in the United States: A National Report on the Racial and Socio-Economic Characteristics of Communities with Hazardous Waste Sites," 1987, xiii.

2. People of Color Environmental Leadership Summit, Principles of Environmental Justice, adopted October 1991.

3. Article two of the United Nations' Convention on the Prevention and Punishment of the Crime of Genocide (passed December 9, 1948)

reads: "In the present Convention, genocide means any of the following acts committed with intent to destroy, in whole or in part, a national, ethnical, racial or religious group, as such: (a) Killing members
of the group; (b) Causing serious bodily or mental harm to members
of the group; (c) Deliberately inflicting on the group conditions of life
calculated to bring about its physical destruction in whole or in part;
(d) Imposing measures intended to prevent births within the group;
(e) Forcibly transferring children of the group to another group."

4. From Bryant's March 1993 testimony before Congress, cited in Christopher Foreman, *The Promise and Peril of Environmental Justice*
(Washington, DC: Brookings Institution, 1998), 75.

5. Daniel Faber, "Green of Another Color: Building Effective Partnerships Between Foundations and the Environmental Justice Movement," Nonprofit Sector Research Fund, April 10, 2001, 6, 42.

6. Our emphasis. Christopher H. Foreman Jr., telephone interview with
M.S., March 23, 2005.

7. Ibid.

8. Foreman, *Promise and Peril*, 3.

9. Ibid.

10. Ibid., 41.

11. Marianne Lavelle and Marcia Coyle, "Unequal Protection," *National
Law Journal*, September 21, 1992, S1–S2.

12. Mary Bryant, "Unequal Justice? Lies, Damn Lies, and Statistics Revisited," *SONREEL News*, September–October 1993, 3–4.

13. Atlas says that Lavelle and Coyle refused to cooperate with his requests
to replicate their study and have offered little in the way of a substan-
tive defense. Mark Atlas, "Rush to Judgment," *Law and Society Review*
35, no. 3, 633–82.

14. Ibid.

15. One of the mistakes made by the *NLJ* reporters was in defining some
zip codes as "minority areas" even if those areas had as few as 20.8 percent people of color in them. "Thus, the penalties against defendants
in those areas do not necessarily reflect what happens to defendants in
predominantly minority areas," Atlas pointed out, "making any comparison to white areas suspect."

Another mistake the *NLJ* reporters made was in categorizing people
who identified themselves on the census as being of Hispanic origin as
whites rather than minorities. Using 1989 census data to construct de-

mographic profiles of the zip codes around facilities, as the *NLJ* did, California goes from being 30.8 percent minority to 42.6 percent if Hispanics are defined as minorities. "Consequently," Atlas concluded, "for all of these reasons the *NLJ* study should never have been considered reliable."

The problems with the *NLJ* study were typical of the problems with environmental justice research generally. It is thus worth quoting from Atlas's statistical detective story at length.

The most serious problem in *NLJ*'s methodology was its failure to control for the effects of penalties on variables other than an area's racial and income characteristics. For example, white area defendants may be more likely to commit environmental violations that traditionally elicit higher penalties; cases in minority areas may be disproportionately concentrated in earlier years, when penalties were lower due to less aggressive enforcement efforts; or high-income area defendants may be more likely to have the financial resources to litigate and cases concluded through litigation, rather than settlement, could result in higher penalties. Essentially, *NLJ*'s approach was, first, to demonstrate that a disparity existed in penalties when penalties were analyzed solely from the perspective of one variable; then, to assume that only that variable caused the disparity; and, finally, to provide the most unsettling explanation for that variable's effect. For example, another possible explanation for lower penalties in minority areas is that, instead of EPA discriminating against people living nearby, more defendants in those areas were minority-run facilities, and EPA thus penalized them less because of EPA's lenient treatment of them.

Atlas, "Rush to Judgment," 2001.

16. Robert D. Bullard, ed., *The Quest for Environmental Justice* (San Francisco: Sierra Club Books, 2005), 31. Likewise, Luke Cole, one of EJ's leading attorneys, and Sheila Foster, a Fordham University law professor, cite the discredited *National Law Journal* findings in their 2001 book *From the Ground Up* (New York: New York University Press, 2001) and fail to mention, much less challenge, the Atlas findings (57). See also Mark Dowie's uncritical embrace of the *National Law Journal* article in *Losing Ground* (Cambridge, MA: MIT Press, 1995), 143–44.

17. Pamela R. Davidson, "Risky Business? Relying on Empirical Studies to Assess Environmental Justice," in *Our Backyard: A Quest for Environmental Justice,* ed. Gerald R. Visgilio and Diana M. Whitelaw (Oxford, England: Rowman and Littlefield, 2003), 91, 98.

18. We could not find examples of environmental justice groups working

on tobacco issues. Richard Drury, a longtime EJ leader with Communities for a Better Environment, confirmed our impression. "I don't know if anyone has done an anti-smoking campaign. It's an important issue. It cuts across communities. But it hasn't traditionally been defined as an environmental issue." Richard Drury, telephone interview with M.S., March 25, 2005.

19. Environmentalist writer Mark Dowie disparaged a Sierra Club staffer who made the obvious point that diet plays a far larger role in creating negative health impacts. "'As surely as Medgar Evers was shot,' said [Sierra Club president Robert] Cox, 'the people of this community are being poisoned.' Not all his colleagues agreed. 'A high fat diet can do it too,' quipped club paralegal Michael Klaes. It was a beginning." Dowie, *Losing Ground*, 169.

20. Harvard Center for Cancer Prevention, "Harvard Report on Cancer Prevention," *Cancer Causes and Control* (1996): 7, cited in Richard Clapp, "Cancer and the Environment," in *Life Support: The Environment and Human Health*, ed. Michael McCally (Cambridge, MA: MIT Press, 2002), 204.

   In 1987, the EPA estimated that pollution caused 1 to 3 percent of all cancers. U.S. Environmental Protection Agency, "Unfinished Business. A Comparative Assessment of Environmental Problems. Appendix I: Report of the Cancer Risk Work Group," Washington, DC, February 1987. Researchers agree with relative certainty that tobacco is responsible for about 30 percent of cancer deaths.

   Most studies conclude that the proportion of cancers due to occupational hazards and air and water pollution is less than 5 percent. Again, public health defines "the environment" as *including* behavior. Interview with Nancy Nelson, and Nancy Nelson, "The Majority of Cancers Are Linked to the Environment," *National Cancer Institute Benchmarks* 4, no. 3 (June 17, 2004).

   As a result, the EPA's National Air Toxics Assessment reports that U.S. residents face a lifetime cancer risk from air pollutants of between 1 and 25 in a million, and a risk of 50 in a million in most urban locations. Alex Kaplun, "EPA Finds Elevated Cancer Rates in Urban Areas," *Greenwire*, February 23, 2006.

   In 1981, British researchers Sir Richard Doll and Sir Richard Peto published a seminal study on the avoidable causes of cancer and concluded that exposure to pollution causes 2 percent of all cancers, food additives another 1 percent, and occupational exposures 4 percent.

Robert Doll and Robert Peto, "The Causes of Cancer: Quantitative Estimates of Avoidable Risks of Cancer in the United States Today," *Journal of the National Cancer Institute* 66 (1981): 1191–1308.

It was criticized as overly conservative by some but has generally held up over time. A follow-up UK study by Doll and Peto resulted in similar estimates. Robert Doll, "Epidemiological Evidence of the Effects of Behavior and the Environment on the Risk of Human Cancer," *Recent Results in Cancer Research* 154 (1998): 3–21.

21. American Lung Association, "Health Disparities in Lung Disease 2006," fact sheet.

22. Centers for Disease Control and Prevention, *Morbidity and Mortality Weekly Report,* May 2004.

23. Luke Cole, telephone interview with M.S., March 15, 2005.

24. Activists whose gaze is not restricted to those things they consider environmental have successfully opposed efforts by tobacco companies to target black Americans. In 1990, R. J. Reynolds was set to market a new menthol cigarette, called Uptown, in mostly black neighborhoods in Philadelphia. Local groups and the American Cancer Society protested the company and persuaded Louis Sullivan, President George H. W. Bush's secretary of health and human services, to publicly criticize Reynolds for promoting a "culture of cancer." Twenty-four hours later the tobacco giant announced it would not introduce Uptown. Described in Foreman, *Promise and Peril,* 118.

25. Robert D. Bullard, "Environmental Justice in the Twenty-first Century," in *Quest for Environmental Justice,* 25.

26. Emphasis in original. Albert L. Nichols, "Risk-Based Priorities and Environmental Justice," in *Worst Things First?* ed. Adam M. Finkel and Dominic Golding (Washington, DC: Resources for the Future, 1995), 268.

27. Bullard erroneously claims that "environmental justice issues were top concerns among African Americans." Robert D. Bullard, ed., *Confronting Environmental Racism: Voices from the Grassroots* (Boston: South End Press, 1993), 7.

28. Paul Mohai, telephone interview with Mark Shellenberger, April 6, 2005.

29. Ibid.

30. In 1993, the CDC found that among children from birth to age four,

blacks were six times more likely to die from asthma than whites; among children ages five to fourteen, blacks were four times more likely to die, and among children ages fifteen to twenty-four, blacks were six times more likely to die. Centers for Disease Control and Prevention, "Asthma Mortality and Hospitalization Among Children and Young Adults — United States, 1980–1993," *Morbidity and Mortality Weekly Report* 45, no. 17: 350–53.

31. L. Rhodes, C. M. Bailey, and J. E. Moorman, "Asthma Prevalence and Control Characteristics by Race/Ethnicity — United States, 2002," *Morbidity and Mortality Weekly Report*, February 27, 2004, 145–48.

32. Richard Pérez-Peña, "Study Finds Asthma in 25 Percent of Children in Central Harlem," *New York Times*, April 19, 2003.

33. Ibid.

34. Eventually 31.4 percent of kids under the age of thirteen were found to be suffering from asthma. Marc Santora, "U.S. Praises Program in City for Children with Asthma," *New York Times*, January 14, 2005.

35. Pérez-Peña, "Study Finds Asthma."

36. Ibid. A similar program focused solely on in-home pollutants dramatically reduced childhood asthma. The National Institutes of Health contracted with seven asthma centers across the country, including one in Harlem. The results were felt almost immediately by the children. In the first year, participating children had twenty-one fewer days of symptoms than nonparticipating children; in the second year, sixteen fewer days. Children improved as quickly as two months after the study began. Mount Sinai School of Medicine, "Mount Sinai Researchers Report Changes in Home Environment Can Reduce Asthma Symptoms in Inner-City Children," press release summary of *New England Journal of Medicine* article, September 8, 2004.

37. Santora, "U.S. Praises Program."

38. Ibid.

39. See also their Web site, www.hcz.org.

40. West Harlem Environmental Action, Inc., "Harlem Asthma Study Confirms WE ACT's Claim: If You Live Uptown, Breathe at Your Own Risk," press release, April 23, 2003.

41. "The first thing that we address, and it continues to be the biggest set

of issues, is the disproportionate rates of morbidity and mortality from asthma and respiratory disease in that specific community. Harlem has the highest asthma mortality rate in the nation." Vernice Miller-Travis, telephone interview with M.S., March 18, 2005.

42. Bullard claims that the environmental justice framework "incorporates the aims of other social movements that seek to eliminate harmful practices in housing (discrimination harms the victim). Land use, industrial planning, health care, and sanitation services. The effects of racial redlining (an illegal practice in which mortgage lenders figuratively draw a red line around minority neighborhoods and refuse to make loans available to those inside the redlined area), economic disinvestment, infrastructure decline, deteriorating housing, lead poisoning, industrial pollution, poverty, and unemployment are not unrelated problems if one lives in an urban ghetto or barrio, in a rural hamlet, or on a reservation." Bullard, *Quest for Environmental Justice*, 25.

  Two things stand out from Bullard's lists. First, the majority of EJ groups do not work on nonpollution issues. WE ACT's fetishization of diesel pollution and its lack of interest in other causes of asthma, including inadequate health care, is typical. Bullard's seemingly expansive framework is simply not borne out by the facts on the ground. Second, it's revealing that Bullard excludes in his definition the most important determinants of human health: diet, exercise, smoking, and alcohol. Such an exclusion is consistent with Bullard's disparagement of risk assessment.

43. On the dangers of fetishizing single causes, like diesel exhaust, Pamela Davidson ("Risky Business," 99) concludes:

  Environmental justice research will have to move beyond a single substance methodology to capture the effects of synergies between chemical substances. The issue of cumulative burden must also be dealt with, since some populations will be exposed to the same substance in multiple settings or will be exposed to substances with synergistic effects. In adopting a wider lens, environmental justice research would benefit from considering other issues pertinent to the spatial configuration of health beyond environmental burdens.

44. The EPA calculates health risk scores for each square kilometer in the

United States using toxic chemical air releases reported by factories. One AP study used these scores to compare regional risks from the long-term exposure to factory pollution. The scores are based on the level of danger to humans posed by each different chemical released, the amount of toxic pollution released by each factory, the path the pollution takes as it spreads through the air, and the number of males and females of different ages who live in the exposure paths. David Pace, "More Blacks Live with Pollution," Associated Press, December 13, 2005.

45. Foreman, *Promise and Peril,* 79–80.
46. Philip Shabecoff, "Environmental Groups Told They Are Racists in Hiring," *New York Times,* February 1, 1990.
47. "The lack of people of color in decision-making positions in your organizations such as executive staff and board positions is also reflective of your histories of racist and exclusionary practices." Southwest Organizing Project Letter, March 15, 1990.
48. Southwest Network for Environmental and Economic Justice letter, May 20, 1990.
49. Robert Gottlieb, *Forcing the Spring* (Washington, DC: Island Press, 1993), 260.
50. Dowie, *Losing Ground,* 152.

4. THE PREJUDICE OF PLACE

1. Robert F. Kennedy Jr., "An Ill Wind off Cape Cod," *New York Times,* December 16, 2005.
2. In his *New York Times* op-ed, Kennedy wrote, "The Humane Society estimates the whirling turbines could every year kill thousands of migrating songbirds and sea ducks."
    The truth is that most environmental groups decided to support the Cape Wind project after extensive studies of wind farms in Europe found very few bird deaths from the slow-moving windmill blades, which most birds easily avoid. See Richard Black, "Wind Farms Pose Low Risk to Birds," *BBC News,* June 8, 2005, and Jack Coleman, "Animals and Wind," *Cape Cod Today,* February 23, 2003.
3. See the fact sheet put out by the National Oceanic and Atmospheric Administration, "Buzzards Bay Oil Spill in Massachusetts," May 2003;

it can be accessed at www.darrp.noaa.gov/northeast/buzzard/pdf/
bbfactsht.pdf.

4. The Clean Power Now fact sheet on health benefits can be accessed
   at www.cleanpowernow.org/images/downloads/Health_1006.pdf. The
   fact sheet was based on Dr. Jonathan Levy and Dr. John D. Spengler,
   "Estimated Public Health Impacts of Criteria Pollutant Air Emissions
   from the Salem Harbor and Brayton Point Power Plants," Harvard
   University School of Public Health, May 2000, 4.

5. Robert F. Kennedy Jr., "Attack Polluters, Not Environmental Leaders,"
   San Francisco Chronicle, January 11, 2006.

6. Bobby Kennedy Jr., "Wind Resistance," *The Connection*, WBUR, October 7, 2002.

7. Carl Pope, "There Is Something Different about Global Warming,"
   response to "The Death of Environmentalism," accessible at www
   .sierraclub.org/pressroom/messages/2004december_pope.asp.

8. Carl Pope, telephone interview with the authors, August 20, 2004.

9. Blossom Hoag, "The Massachusetts Chapter's Position on the Cape
   Cod Wind Farm," *Massachusetts Sierran,* Summer 2005, 3.

10. Unidentified caller, Kennedy, "Wind Resistance."

11. Kevin Dennehy, "Our Future? General Electric Has Invested Millions
    in This Irish Offshore Wind Complex. Nantucket Sound Could Be
    Next," *Cape Cod Times,* November 6, 2005.

5. The Pollution Paradigm

1. Al Gore, *An Inconvenient Truth* (Emmaus, PA: Rodale, 2006), 12.

2. Ibid., 109.

3. Internet Movie Database, www.imdb.com/title/tt0497116/taglines.

4. "Be Worried. Be Very Worried." *Time*, March 26, 2006.

5. Katherine Ellison, "Turned Off by Global Warming," *New York Times*,
   May 20, 2006.

6. Gore, *Inconvenient*, 286.

7. Emphasis ours. From the film version of *An Inconvenient Truth*. Also
   quoted in David Neff, "Al Gore, Preacher Man," *Christianity Today*,
   May 31, 2006.

8. The Pew Center is funded by the same foundation that invests tens
   of millions annually into environmental causes. A 2001 *New York*

Times article named Pew as the largest grant maker to environmental causes: $51 million in environmental causes in 2000 alone. Douglas Jehl, "Charity Is New Force in Environmental Fight," *New York Times,* June 28, 2001.

9. Pew Research Center for the People and the Press, "Partisanship Drives Opinion: Little Consensus on Global Warming," July 12, 2006, accessible at www.people-press.org.

10. Pew Research Center for the People and the Press, "Global Warming: A Divide on Causes and Solutions: Public Views Unchanged by Unusual Weather," January 24, 2007.

11. Gore, *Inconvenient,* 9. Gore neglects to mention that the Democrats actually held both houses of Congress for the first two years of the Clinton presidency — the period in which presidents tend to achieve much of their agendas (this was the period President George W. Bush pushed through his tax cuts).

12. Thomas Kuhn, *The Structure of Scientific Revolutions* (Chicago: Lawrence Van Gelder, 1962); "Obituary: Thomas Kuhn, 73; Devised Science Paradigm," *New York Times,* June 19, 1996.

13. This point was eloquently made in Jeffrey L. Kasser's cogent lectures on the philosophy and history of science in *The Philosophy of Science: Kuhn and the Challenge of History,* available on DVD from the Teaching Company, 2006.

14. Kuhn, *Structure,* 92.

15. Max Planck, *Scientific Autobiography and Other Papers* (New York: Philosophical Library, 1949).

16. Carl Pope, "Response to 'The Death of Environmentalism': There Is Something Different about Global Warming," December 2004, www.sierraclub.org/pressroom/messages/2004december_pope.asp.

17. Phil Clapp, "Over Our Dead Bodies: Green Leaders Say Rumors of Environmentalism's Death Are Greatly Exaggerated," interview by Amanda Griscom Little, January 13, 2005, accessible at www.grist.org.

18. Associated Press, "Global Warming Gases on Rise Again Despite Kyoto Protocol," October 30, 2006; "Narrow Path Between Protecting Trade and Saving the Planet," Editorial, *New York Daily News,* November 6, 2006.

19. United Nations, Framework Convention on Climate Change, "Changes in GHG emissions from 1990 to 2004 for Annex I Parties," http://unfccc.int/files/essential_background/background_publications

_htmlpdf/application/pdf/ghg_table_06.pdf; Stephen Leahy, "Kyoto Gets a Slap in Face from Canada," Inter Press Service News Agency, December 9, 2006.

20. "Selling Hot Air," *Economist*, September 9, 2006.

21. Alain Bernard, Sergey Paltsev, John M. Reilly, Marc Vielle, and Laurent Viguier, "Russia's Role in the Kyoto Protocol," *MIT Joint Program on the Science and Policy of Global Change*, report no. 98, June 2003.

22. In 2001, the United States emitted 5.5 metric tons of carbon per person, while China emitted 0.6 tons per person. U.S. Energy Information Administration, "China: Environmental Issues," fact sheet, www.eia.doe.gov/emeu/cabs/chinaenv.pdf.

23. Richard Balzhiser, "The Chinese Energy Outlook," *National Academy of Engineering* 28, no. 2 (Summer 1998); Keith Bradsher and David Barboza, "Pollution from Chinese Coal Casts a Global Shadow," *New York Times*, June 11, 2006.

24. Rob Collier, "A Warming World: China about to Pass U.S. as World's Top Generator of Greenhouse Gases," *San Francisco Chronicle*, May 5, 2007.

25. Keith Bradsher, "China to Pass U.S. in 2009 in Emissions," *New York Times*, November 7, 2006; Robert Samuelson, "The Real Inconvenient Truth," *Newsweek*, July 5, 2006. Samuelson writes, "From 2003 to 2050, the world's population is projected to grow from 6.4 billion people to 9.1 billion, a 42 percent increase. If energy use per person and technology remain the same, total energy use and greenhouse gas emissions (mainly carbon dioxide) will be 42 percent higher in 2050. But that's too low, because societies that grow richer use more energy. Unless we condemn the world's poor to their present poverty — and freeze everyone else's living standards — we need economic growth. With modest growth, energy use and greenhouse emissions more than double by 2050."

26. This point was made to us by Professor Dara O'Rourke of UC Berkeley in an interview on August 2, 2006. See also "China: Harsh Sentences for Labor Activists," *Human Rights Watch*, May 10, 2003, and Han Dongfang, "Chinese Labor Struggles," *New Left Review*, July–August 2005.

27. Lu Xuedu, deputy director general of the Chinese Office of Global Environmental Affairs, in Bradsher, "China to Pass U.S."

28. Jim Yardley, "China Says Rich Countries Should Take Lead on Global Warming," *New York Times*, February 7, 2007.

29. Wan Zhihong, "NDRC Announces Coal Chemical Plans," *China Daily*, December 28, 2006. In December 2006 the government spent $639 million to cover the losses of the state oil company Sinopec. Wang Yu, "Sinopec Given One-Off State Subsidy of US$639 m," *China Daily*, December 28, 2006.

30. *China Daily*, December 18, 2006.

31. Nicholas Stern, "Stern Review on the Economics of Climate Change," October 30, 2006, accessible at www.hm-treasury.gov.uk/ independent_reviews/stern_review_economics_climate_change/stern review_index.cfm. Stern quoted from Robert Peston, "Report's Stark Warning on Climate Change," *BBC News*, October 29, 2006.

32. In the weeks following the release of the Stern Review, Stern was criticized for erroneously setting the value of future consumption as equal to the value of present-day consumption. The criticism was important, because there is a reasonable case to be made that economies, especially in the developing world, will and should consume more today, in order to make the climb out of poverty through investments in such things as basic infrastructure, than in the future.

33. See www.ipcc.ch.

34. See, for instance, Daniel Kammen, "The Rise of Renewable Energy," *Scientific American*, September 2006, 84–93; W. Wayt Gibbs, "Plan B for Energy," *Scientific American*, September 2006, 102–14; Joan Ogden, "High Hopes for Hydrogen," *Scientific American*, September 2006, 94–101; Eberhard Joachem, "An Efficient Solution," *Scientific American*, September 2006, 64–67; Robert H. Socolow and Stephen H. Pacala, "A Plan to Keep Carbon in Check," *Scientific American*, September 2006, 50–59.

35. See, for instance, Michael Grubb, "Technology Innovation and Climate Change Policy: An Overview of Issues and Options," www.econ .cam.ac.uk/faculty/grubb/publications/J38.pdf, and D. Sarewitz and R. A. Pielke Jr., "The Steps Not Yet Taken," in *Controversies in Science and Technology*, vol. 2, *From Chromosomes to the Cosmos*, ed. Daniel Lee Kleinman, Karen Cloud-Hansen, Christina Matta, and Jo Handelsman (New Rochelle, NY: Mary Ann Liebert, Inc., 2007).

36. Editorial, "Cooling Our Heels," *Scientific American*, September 2006, 8.

37. Stern, "Stern Review," 347–65.

38. Ibid., 367.

39. Kammen, "The Rise of Renewable Energy," 87.

40. Ibid., 92.

41. Ibid.

42. In the words of energy policy experts Daniel Sarewitz and Roger Pielke Jr., clean-energy technologies "can overcome political conflict not by compelling diverse interests and values to converge — the job assigned to climate science — but by allowing them to co-exist in a shared sense of practical benefits." Sarewitz and Pielke, "The Steps Not Yet Taken."

43. Thoreau, *Walden*, 31, 38.

44. Amanda Griscom Little, "For a Moment: John Kerry and Teresa Heinz Kerry Chat about Their New Environmental Book," March 29, 2007, accessible at www.grist.org.

45. Gore, *Inconvenient*, 286.

46. Anne Paine, "Group Questions Level of Energy Use at Gore Home," *Tennessean*, February 27, 2007.

47. "Gore also owns homes in Carthage, Tenn., and in the Washington area." Kristen Hall, "Conservative Group Calls Gore Hypocrite," Associated Press, February 28, 2007.

48. Gore, *Inconvenient*, 220.

49. Ibid., 218–19.

50. "Imagine for a moment a world where cities have become peaceful and serene because cars and buses are whisper quiet, vehicles exhaust only water vapor, and parks and greenways have replaced unneeded urban freeways. OPEC has ceased to function because the price of oil has fallen to five dollars a barrel, but there are few buyers for it because cheaper and better ways now exist to get the services people once turned to oil to provide. Living standards for all people have dramatically improved, particularly for the poor and those in developing countries. Involuntary unemployment no longer exists, and income taxes have largely been eliminated. Houses, even low-income housing units, can pay part of their mortgage costs by the energy they *produce;* there are few if any active landfills; worldwide forest cover is increasing; dams are being dismantled; atmospheric CO_2 levels are decreasing for the first time in two hundred years; and effluent water leaving factories is cleaner than the water coming into them. Industrialized countries have reduced resource use by 80 percent while improving the quality of life . . . Is this the vision of utopia? In fact, the changes described here could come about in the decades to come as a result of economic and technological trends already in place." Paul Hawken,

Amory Lovins, and L. Hunter Lovins, *Natural Capitalism: Creating the Next Industrial Revolution* (New York: Little, Brown, 1999), 1.

51. Bill McKibben, "Changing the Climate: Why a New Approach to Global Warming Would Make for a Better Politics — and Planet," *American Prospect*, October 2005, A10.

52. Bill McKibben, "What 'Green' Means," *Philadelphia Inquirer*, August 7, 2005; originally published in *Orion*, July–August 2005.

6. THE DEATH OF ENVIRONMENTALISM

1. The words belong not to her but to the man to whom she dedicated the book, Albert Schweitzer.

2. T. S. Eliot, "Quartet No. 1: Burnt Norton," *Four Quartets* (Orlando, FL: Harcourt, 1943, 1968), 2.

3. C. G. Jung, *Mysterium Coniunctionis: An Inquiry into the Separation and Synthesis of Psychic Opposites in Alchemy* (New York: Pantheon, 1963).

4. Rachel Carson, *Silent Spring* (New York: Houghton Mifflin, 1962), 2–3.

5. "Nature," Carson writes, sets up a "wonderful and intricate system of checks and balances." *Silent Spring*, 293. "Ecology is the study of balance," wrote Al Gore, *Earth in the Balance* (New York: Houghton Mifflin, 1992), 11. "Why, after millions of years of harmonious co-existence, have the relationships between living things and their earthly surroundings begun to collapse?" Barry Commoner, *The Closing Circle* (New York: Knopf, 1971), 8.

6. "Change now appears to be intrinsic and natural at many scales of time and space in the biosphere. Wherever we seek to find constancy . . . we discover change." Daniel Botkin, *Discordant Harmonies* (New York: Oxford University Press, 1990), 9.

7. Charles C. Mann, *1491: New Revelations of the Americas Before Columbus* (New York: Knopf, 2005).

8. Carson, *Silent Spring*, 297.

9. René Dubos, *So Human an Animal* (New York: Scribner, 1968), 7.

10. Even the language of connection between nature and humans, the geographer Jim Proctor has noted, presumes a separation. "Greater connection is not, then, needed between people and the environment," he writes, "but rather the deconstruction of both categories. Jim

Proctor, "Environment After Nature," 2007, 8, accessible at www
.lclark.edu/~jproctor/pdf/NV2007Draft.pdf.

11. Jared Diamond, *Collapse* (New York: Viking, 2005), 230–33.
12. Ibid., 431.
13. Michael Oppenheimer, e-mail May 10, 2005.
14. Michael Crichton, "Environmentalism as Religion," Commonwealth Club, San Francisco, September 15, 2003.
15. Michael Crichton, *State of Fear* (New York: Harper, 2004), 637–38.
16. Thomas A. Sancton, "What on Earth Are We Doing?" *Time*, January 2, 1989.
17. Emphasis ours. Ibid., 48.
18. Thomas Berry, *The Dream of the Earth* (San Francisco: Sierra Club Books, 1990).
19. Dubos is a partial exception. Though he elevated an antiquated notion of nature as harmonious and spoke of our collective guilt, he also recognized the significance of changing values. "The most hopeful sign for the future is the attempt by the rebellious young to reject our social values . . . As long as there are rebels in our midst there is reason to hope that our societies can be saved." Dubos, *So Human*, 5.
20. Edward O. Wilson, *The Future of Life* (New York: Random House, 2002), 214.
21. William Chaloupka, "Jagged Terrain: Cronon, Soulé, and the Struggle over Nature and Deconstruction in Environmental Theory," *Strategies* 13, no. 1 (2000): 29.
22. Julia Hill, the daughter of a preacher, tells a Christian story about being chosen by God ("the Great Spirit"), who had answered her prayers and "was working miracles in my life," and says that "Luna spoke to me and reminded me of something I had received in prayer." Julia Butterfly Hill, *The Legacy of Luna* (New York: Harper San Francisco, 2000), 133, 246.
23. John Gray, *Straw Dogs* (London: Granta, 2002), 17.
24. Carl Pope, "There Is Something Different about Global Warming," December 14, 2004.
25. "Talking Primates with Dr. Frans de Waal," August 25, 2005, http://blog.92y.org/index.php/item/dr_frans_de_waal/.
26. Gray, *Straw Dogs*, 14–15.
27. Ibid., 17.

28. Diamond, *Collapse,* 495–96.

29. "One of those choices has depended on the courage to practice long-term thinking . . . The other crucial choice illuminated by the past involves the courage to make painful decisions about values." Ibid., 522–23.

30. During these experiments, social psychologists prompt subjects to think about their own deaths, and then the experimenters examine the subjects' behavior and attitudes. What they find in these death-priming exercises is not generosity and altruism but the propensity to scapegoat out-group members (for example, racial, religious, and sexual minorities), justify inequality and authoritarianism, and resist change (system-justify). The authors of a major research review summarized this research in 2003.

    By leading experimental research participants to anticipate the cognitive and affective experience of death (e.g., Rosenblatt et al., 1989), they have demonstrated that mortality salience leads people to defend culturally valued norms and practices to a stronger degree (Greenberg et al., 1990, 1995) and to distance themselves from, and even to derogate, out-group members to a greater extent (Harmon-Jones, Greenberg, Solomon, & Simon, 1996; McGregor et al., 2001). In addition, the fear of death has been linked to system-justifying forms of stereotyping and enhanced liking for stereotype-consistent women and minority group members (Schimel et al., 1999).

    John T. Jost et al., "Political Conservatism as Motivated Social Cognition," *Psychological Bulletin* 129, no. 3 (2003): 365.

31. Diamond, *Collapse,* 558.

32. "Under conditions of heightened mortality salience, defense and justification of the worldview should be intensified, thereby decreasing tolerance of opposing views and social, cultural, and political alternatives," wrote Jost et al. "Political Conservatism," 349.

33. Ibid.

34. The word *multinaturalism* is Bruno Latour's. See *Politics of Nature: How to Bring the Sciences into Democracy* (Cambridge, MA: Harvard University Press, 2004), 29.

35. Richard Rorty, *Contingency, Irony, and Solidarity* (Cambridge, England: Cambridge University Press, 1989), 7.

36. Jane Bennett, *The Enchantment of Modern Life* (Princeton, NJ: Princeton University Press, 2001), 4.

7. STATUS AND SECURITY

1. Thomas Frank, *What's the Matter with Kansas?: How Conservatives Won the Heart of America* (New York: Henry Holt, 2004).

2. Ibid., 7.

3. Ibid., 1.

4. This is the average median income for Kansas for the years 2002, 2003, and 2004. U.S. Bureau of the Census, "Three-Year-Average Median Household Income by State: 2002–2004, in 2004 Dollars," Washington, DC, 2004.

5. In 2001, its unemployment rate was 4.3 percent, compared with the U.S. rate of 4.8 percent. In 2003, the poverty rate in Kansas was 10.4 percent, compared with 12.5 percent in the United States. U.S. Bureau of the Census, www.fedstats.net/qf/states/20000.html.

6. Life-expectancy rates in Kansas are slightly above the U.S. average. A Kansas resident born in 2000 had a life expectancy of 77.4 years, compared to a U.S. life expectancy of 76.9 years. Kansas Department of Health and Environment, "Kansas Abridged Life Tables, 2000," May 2003, www.kdhe.state.ks.us/ches/liftab00.pdf.

7. Less than 20 percent of Kansans smoke, which ranks them as eleventh best in the country. United Health Foundation, "America's Health Rankings: 2005 Edition," www.unitedhealthfoundation.org/shr2005/states/Kansas.html.

8. 69.2 percent of Kansans are homeowners, which is higher than the U.S. average of 66.2 percent. U.S. Bureau of the Census, www.fedstats.net/qf/states/20000.html.

9. Only 11.1 percent of Kansans are without health insurance, which ranks them sixth best in the country. United Health Foundation, "America's Health Rankings."

10. In 2000, 86 percent of Kansans over age 25 were high school graduates, as compared to 80.4 percent of the U.S. population. U.S. Bureau of the Census, www.fedstats.net/qf/states/20000.html.

11. Ibid. In 2000, 25.8 percent of Kansans over age twenty-five held a bachelor's degree or higher, compared to 24.4 percent of the U.S. population.

12. Ibid. The average commute time (one way) for Kansas workers sixteen and older is nineteen minutes, which is lower than the U.S. average commute of 25.5 minutes.

13. Frank, *What's the Matter,* 104.
14. Benjamin M. Friedman, *The Moral Consequences of Economic Growth* (New York: Knopf, 2005), 208–9.
15. The *American Prospect*'s Garance Franke-Ruta made a similar point in her December 2004 article "Health Thy 'Self'":

    Self-interest defined in purely economic terms is an idea that reduces the Democratic Party to little more than the human resources department of American politics, endlessly fussing over pensions and health-care plans and whether or not you got your flu shot, rather than a party concerned with the fundamental stuff of life: who we are, how we organize our society, and what it means to be an American at this particular moment in history. Garance Franke-Ruta, "Health Thy 'Self,'" *American Prospect*, December 2004, 34.

16. Harvey Levenstein, *A Paradox of Plenty: A Social History of Eating in Modern America* (Berkeley: University of California Press, 2003), 3–4.
17. Friedman, *Moral Consequences,* 181–82.
18. Ibid., 183–84.
19. Ibid., 185.
20. In 1946, President Truman created the first federal Civil Rights Commission and extended the Fair Employment Practices Committee. In 1947, the Dodgers' Jackie Robinson broke the race barrier in baseball. A year later, Truman proposed federal action against lynching and discrimination, and the Supreme Court invalidated homeowners' covenants that maintained religious and racial segregation. By 1953, the Court had banned white-only primary voting common among southern political parties, and shortly thereafter the NAACP declared the "virtual disappearance" of lynching.

    Young blacks graduating from high school went from 24 percent in 1950 (compared to 56 percent of whites) to 77 percent in 1980 (compared to 89 percent of whites), and the average reading proficiency of black seventeen-year-olds went from 82 percent of the average score for whites in 1971 to 93 percent in 1988. Friedman, *Moral Consequences,* 186–89.
21. Ibid., 188–89.
22. Ibid., 192–94.
23. Ibid., 197–98.
24. Ibid., 198–99.
25. Juliet Schor, *The Overspent American* (New York: Basic Books, 1998), 12.
26. Friedman, *Moral Consequences,* 199, 200, 209.

27. Ibid., 208.
28. Environics, Inc., "3SC Survey, 1992–2004," www.americanenvironics.com. Based on a 2,500-person in-home survey fielded by Roper.
29. Michael Adams, *American Backlash: The Untold Story of Social Change in the United States* (New York: Penguin, 2005), 37–39.
30. Cited in Greg Critser, *Fat Land: How Americans Became the Fattest People in the World* (Boston: Houghton Mifflin, 2003), 117.
31. According to the Centers for Disease Control and Prevention, one-third of Americans born in 2001 will develop diabetes in their lifetimes. Cited in N. R. Kleinfield, "Diabetes and Its Awful Toll Quietly Emerge as a Crisis," *New York Times,* January 9, 2006.
32. Kentucky, one of America's poorest states, spends more than $1 billion annually on obesity-related medical expenses. Laura Ungar, "Kentucky Health: Critical Condition; Bad Habits, Poverty Undermine Health," *Louisville Courier-Journal,* July 17, 2005.
33. Critser, *Fat Land,* 111.
34. Jacob Hacker, *The Great Risk Shift* (Oxford, England: Oxford University Press, 2006), 13, 27.
35. Roper Center study, University of Connecticut, published in *American Enterprise,* May–June 1993, 87, and cited in Schor, *Overspent American,* 16.
36. "Most people understandably exhibit generosity when they are doing well and defensiveness when they are doing badly. But they pull together when they see their very lives threatened and the entire social and political structure in which they live thrown into imminent danger." Friedman, *Moral Consequences,* 178.
37. Schor, *Overspent American,* 4.
38. Ibid., 6.
39. Ibid., 13.
40. Ibid., 19–20.
41. Steve Lohr, "Ka-Ching: Maybe It's Not All Your Fault," *New York Times,* December 5, 2004.
42. Schor, *Overspent American,* 19.
43. Roger Cohen, "Communists as Creditors: China and the U.S. Worker," *New York Times,* May 22, 2005.
44. Personal savings rate reached minus .05 percent in 2005. Associated Press, "Consumer Debt Rises, But Less Than Expected," *New York Times,* June 8, 2005.
45. Ibid.

46. Vikas Bajaj and Ron Nixon, "For Minorities, Signs of Trouble in Fore-closures," *New York Times,* February 22, 2006.

47. Niall Ferguson, "Reasons to Worry," *New York Times Magazine,* June 11, 2006, 50.

48. Walter Kirn, "Home Sweet Debt," *New York Times,* March 5, 2006.

49. Our focus group research was done in 2006 with Celinda Lake and David Mermin of Lake Research Partners for Dr. Robert Crittenden of the Herndon Alliance, which was funded by the Nathan Cummings Foundation, the Public Welfare Foundation, and the California Endowment.

50. Frank, *What's the Matter,* 237.

51. Frank, "What's the Matter with Liberals?," *New York Review,* May 12, 2005, 48.

52. Frank, *What's the Matter,* 251.

8. Belonging and Fulfillment

1. Thomas Frank, *What's the Matter with Kansas?: How Conservatives Won the Heart of America* (New York: Henry Holt, 2004), 168.

2. Jeremy Kahn, "Will These People Swing?," *Fortune,* March 22, 2004.

3. From Matt Bai, "Who Lost Ohio?," *New York Times Magazine,* November 21, 2004:

> In shiny new town-house communities, canvassing could be done quietly by neighbors; you didn't need vans and pagers. Polling places could accommodate all the voters in a precinct without ever giving the appearance of being overrun. In the old days, these towns and counties had been nothing but little pockets of voters, and Republicans hadn't bothered to expend the energy to organize them. But now the exurban populations had reached critical mass (Delaware County alone had grown by almost one-third since the 2000 election), and Republicans were building their own kind of quiet but ruthlessly efficient turnout machine . . .
>
> But with so many white, conservative and religious voters now living in the brand-new town houses and McMansions in Ohio's growing ring counties, Republicans were able to mobilize a stunning turnout in areas where their support was more concentrated than it was in the past. Bush's operatives did precisely what they told me seven months ago they would do in these communities: they tapped into a volunteer network using local party organizations, union rolls, gun clubs and churches.

4. Michelle Goldberg, *Kingdom Coming* (New York: W. W. Norton, 2006), 21, 58.

5. "Indeed, even close observers of Ohio politics might have missed the Bush campaign's emphasis on social values because much of its outreach efforts occurred away from the mass media," Paul Farhi and James V. Grimaldi noted in "GOP Won with Accent on Rural and Traditional," *Washington Post,* November 4, 2004. See also Darrel Rowland, "How Bush Pulled It Off in Ohio," *Columbus Dispatch,* May 15, 2005, and Goldberg, *Kingdom Coming,* 57–69.

6. Goldberg, *Kingdom Coming,* 21.

7. Ibid., 181.

8. Ibid., 191.

9. Isaac Kramnick, introduction to *Democracy in America,* by Alexis de Tocqueville (New York: Penguin, 2003), xx; Herbert Marcuse, *One-Dimensional Man* (New York: Beacon, 1964, 1991).

10. Tocqueville, *Democracy,* 596.

11. Robert D. Putnam, *Bowling Alone* (New York: Simon and Schuster, 2000).

12. Miller McPherson, Lynn Smith-Lovin, and Matthew E. Brashears, "Social Isolation in America: Changes in Core Discussion Networks over Two Decades," *American Sociological Review,* June 2006, 353–73.

13. Richard Florida, *The Rise of the Creative Class* (New York: Perseus, 2002), 8.

14. Ibid., 269.

15. Mark Granovetter, "The Strength of Weak Ties," *American Journal of Sociology,* May 1973, 1360–80. Also see discussion in Florida, *Rise,* 276.

16. Florida, *Rise,* 277.

17. Ibid., 281.

18. Ibid., 269.

19. Malcolm Gladwell, "The Cellular Church," *The New Yorker,* September 12, 2005, 63.

20. Putnam, *Bowling,* 129.

21. Gladwell, "Cellular Church," 63.

22. Staff article, "Rick Warren, Jerry Falwell Meet at 'Purpose-Driven' Common Ground with 13,000," Pastors.com, 2003.

23. Robert Putnam and Lewis Feldstein with Don Cohen, *Better Together: Restoring the American Community* (New York: Simon and Schuster, 2003), 133.

24. Tocqueville, *Democracy*, 598.

25. Gladwell, "Cellular Church," 67.

26. American Environics, "Toward a New Ecological Majority," December 2006, http://americanenvironics.com.

27. Steven Rose, "What's Not the Matter with the Middle Class?," *American Prospect*, September 4, 2006.

28. See Jonathan Haidt, *The Happiness Hypothesis* (New York: Basic, 2006); Daniel Gilbert, *Stumbling on Happiness* (New York: Knopf, 2006); Martin Seligman, *Authentic Happiness* (New York: Simon and Schuster, 2002); Robert Lane, *The Loss of Happiness in Market Democracies* (New Haven: Yale University Press, 2000); Timothy Wilson, *Strangers to Ourselves* (Cambridge, MA: Harvard University Press, 2002); and Gregg Easterbrook, *The Progress Paradox* (New York: Random House, 2003).

29. See, for instance, James Banks et al., "Disease and Disadvantage in the United States and in England," *Journal of the American Medical Association*, May 3, 2006.

30. Tocqueville, *Democracy*, 600.

31. Richard Cizik, "To the Point: Evangelicals and the Environment," *Warren Olney Show*, Public Radio International, February 9, 2005.

32. See Susan Friend Harding, *The Book of Jerry Falwell: Fundamentalist Language and Politics* (Princeton, NJ: Princeton University Press, 2000).

33. Richard Rorty, *Contingency, Irony, and Solidarity* (Cambridge, England: Cambridge University Press, 1989), xiv.

34. Richard Rorty, *Achieving Our Country* (Cambridge, MA: Harvard University Press, 1998), 101.

35. Ibid., 34.

36. "There is a time in every man's education when he arrives at the conviction that envy is ignorance; that imitation is suicide; that he must take himself for better, for worse, as his portion; that though the wide universe is full of good, no kernel of nourishing corn can come to him but rough his toil be sowed on that plot of ground which is given him to till . . . Your goodness must have some edge to it — else it is none . . . Your conformity explains nothing. Act singly, and what you have already done singly will justify you now. Greatness appeals to the future." Ralph Waldo Emerson, "Self-Reliance," in *The Essential Writings of Ralph Waldo Emerson* (New York: Random House, 2000), 133, 135, 139.

37. "Become who you are!" Friedrich Nietzsche, *Thus Spoke Zarathustra* (New York: Barnes and Noble Books, 2005).

38. Abraham Maslow, *Toward a Psychology of Being* (New York: John Wiley and Sons, 1968, 1999), 149–234.

39. "If we value independence, if we are disturbed by the growing conformity of knowledge, of values, of attitudes, which our present system induces, then we may wish to set up conditions of learning which make for uniqueness, for self-direction, and for self-initiated learning." Carl Rogers, *On Becoming a Person* (Boston: Houghton Mifflin, 1961, 1995).

40. For a discussion of self-expression values, see Ronald Inglehart and Christian Welzel, *Modernization, Cultural Change, and Democracy* (Cambridge, England: Cambridge University Press, 2005), 1, 5: "Emerging self-expression values transform modernization into a process of human development, giving rise to a new type of humanistic society that is increasingly people-centered . . . In the postindustrial phase of modernization, another cultural change becomes dominant — a shift from survival values to self-expression values, which brings increasing emancipation *from* authority. Rising self-expression values transform modernization into a process of human development that increases human freedom and choice."

41. "The thing is to find a truth which is true for me, to find the idea for which I can live and die," Søren Kierkegaard, *The Journals of Søren Kierkegaard,* ed. Alexander Dru (Oxford, England: Oxford University Press, 1938).

42. Martin Heidegger, *Being and Time* (Albany: State University of New York, 1996), 213–306.

43. *Regrounding* is a term used by the social-values research firm Sociovision. See www.socio-vision.com.
     Sociovision researchers found that among Europeans, there has been a re-gendering in the form of "women wanting to feel more 'womanly' and men wanting to feel more 'manly.' This does not herald a return to Traditional Role Model but has more to do with finding meaning and a sense of identity." The researchers also found evidence that "gender moods" — various expressions of femininity and masculinity — are exercised in different settings, in shopping, sports, work, and entertainment. Sociovision, "The Drivers of Change in a Cross-Cultural World," *GlocalConsult,* 2002.

9. PRAGMATISM

1. Over the last forty years, cognitive therapy has been studied in multiple large clinical studies and has been found to be as effective as, or more effective than, drugs like Prozac; unlike psychotropic drugs, cognitive therapy's benefits continue long after it stops. Our point is not that cognitive therapy is superior to drugs or to other therapies but that it is a powerful case study showing how narratives shape our emotional lives. See Jonathan Haidt, *The Happiness Hypothesis* (New York: Basic Books, 2006), 37–39.

2. Louis Menand, *The Metaphysical Club* (New York: Farrar, Straus and Geroux, 2001), 236.

3. Al Gore, *Earth in the Balance* (New York: Houghton Mifflin, 1992), 240; Andrew Revkin, "Climate Talks Shift Focus to How to Deal with Changes," *New York Times*, November 3, 2002; Carl Pope, "Response to the Death of Environmentalism: There Is Something Different about Global Warming," December 2004.

4. Carl Pope, "Preparing for Global Warming," *San Francisco Chronicle*, October 2, 2005.

5. We have not been alone in our advocacy of adaptation and preparedness. In 2000, science policy analysts Daniel Sarewitz and Roger Pielke Jr. laid out a similar approach in "Breaking the Global Warming Gridlock," *Atlantic Monthly*, July 2000, 54–64.

6. Dan Becker, telephone interview with authors, September 15, 2004.

7. Richard Rorty, *Achieving Our Country* (Cambridge, MA: Harvard University Press, 1998), 107.

8. If we defined a species only as, say, those organisms that can reproduce with one another, we wouldn't be able to categorize species that reproduce asexually. Defining a species as an organism's evolutionary lineage solves the problem posed by the reproduction-based definition, but it faces the problem of time. Where do we draw the temporal boundaries between species in the process of evolution? The question itself demolishes the notion of natural kinds of categories existing separately from humans.

   Our species is a case in point. When was that magic moment when we broke away from other apes to become humans? Scientists have long tried to reduce this process of becoming humans to particular

moments — the use of tools, the birth of language, and so forth. But
there is no defining characteristic, just as there was no magic moment.
Rather, there were infinite moments, albeit ones punctuated at particular instants with productive kinds of evolutionary chaos, including
climate change.

Different definitions of species have been crucial for biologists to
accomplish miraculous things. Efforts to create a single definition
would be unscientific. A more pragmatic view would encourage different uses of the term *species* within the various biological sciences.
The physicist and historian of science Paul Feyerabend observed, The
result that emerged in the nineteenth century was not a coherent science but a collection of heterogeneous subjects (optics, acoustics, hydrodynamics, elasticity, electricity, heat, etc. in physics; physiology,
anatomy, etc. in medicine; morphology, evolutionism, etc. in biology
— and so on). Some of the subjects (hydrodynamics, for example)
had only a tenuous relation to experiment, others were crudely empirical. It did not matter. Being firmly convinced that the world was uniform and subjected to 'inexorable and immutable laws' leading scientists interpreted the collection as an *appearance* concealing a uniform
material *reality.*

And Feyerabend writes: "The idea assumes that the world is divided
into (at least) two regions — a primary region consisting of important events and a secondary region that differs from the primary region, blocks our vision, is deceptive and, in many cases, evil. Grand di-
chotomies of this kind are found in many though by no means in all
cultures." Paul Feyerabend, *Conquest of Abundance: A Tale of Abstraction Versus the Richness of Being* (Chicago: University of Chicago Press,
1999), 167.

9. Bruno Latour makes a similar point in *The Politics of Nature: How to
Bring the Sciences into Democracy* (Cambridge, MA: Harvard University Press, 2004), 9–10, 24. "In all that follows, I shall ask my readers to
agree to dissociate *the sciences* — in the plural and in small letters —
from *Science* — in the singular and capitalized. I ask readers to acknowledge that discourse on Science has no direct relation to the life
of the sciences, and that the problem of knowledge is posed quite differently, depending on whether one is brandishing Science or clinging
to the twists and turns of the sciences as they are developed. I ask readers finally to grant that if nature — in the singular — is closely linked
with Science, the sciences for their part in no way require such unifica-

tion. If we were trying to approach the question of political ecology as if Science and the sciences were one and the same enterprise, we would end up in radically different positions . . . I am going to define Science as the *politicization of the sciences through epistemology in order to render ordinary political life impotent through the threat of an incontestable nature.*"

And later, "Replace the singular with the plural everywhere. Suddenly we have *natures,* and it is impossible to make natures play any political role whatsoever."

10. GREATNESS

1. Ron Suskind, "Without a Doubt," *New York Times Magazine,* October 17, 2004.
2. John T. Jost et al., "Political Conservatism as Motivated Social Cognition," *Psychological Bulletin* 129, no. 3 (2003): 339–75.
3. Francis Fukuyama, *America at the Crossroads: Democracy, Power, and the Neoconservative Legacy* (New Haven: Yale University Press, 2006).
4. The January 26, 1998, letter to President Clinton that Fukuyama signed reads, "The only acceptable strategy is one that eliminates the possibility that Iraq will be able to use or threaten to use weapons of mass destruction. In the near term, this means a willingness to undertake military action as diplomacy is clearly failing. In the long term, it means removing Saddam Hussein and his regime from power. That now needs to become the aim of American foreign policy."

    The letter can be read online at www.newamericancentury.org/iraqclintonletter.htm.
5. The September 20, 2001, letter to George W. Bush that Fukuyama signed reads:

    We agree with Secretary of State Powell's recent statement that Saddam Hussein "is one of the leading terrorists on the face of the Earth." It may be that the Iraqi government provided assistance in some form to the recent attack on the United States. But even if evidence does not link Iraq directly to the attack, any strategy aiming at the eradication of terrorism and its sponsors must include a determined effort to remove Saddam Hussein from power in Iraq. Failure to undertake such an effort will constitute an early and perhaps decisive surrender in the war on international terrorism. The United States must therefore provide full military and financial support to the Iraqi opposition. American military force should be used to provide a "safe zone" in Iraq from which the opposition can operate. And American forces must be prepared to back up our commitment to the Iraqi opposition by all necessary means.

The full letter can be read online at www.newamericancentury.org/ Bushletter.htm.

6. Fukuyama tells a story of himself as an independent thinker brave enough to break with his fellow neoconservatives over the war in Iraq. "I was never persuaded of the rationale for the Iraq war," he writes in *America at the Crossroads*.

   Fukuyama acknowledges that he signed the 1998 letter to Clinton but then adds, in order to downplay the significance of the letter, "An American invasion of Iraq was not then in the cards, however, and would not be until the events of September 11, 2001" (x). Fukuyama implies that his position changed after 9/11, but if that were the case then he would not have signed the September 20, 2001, letter urging President George W. Bush to remove Hussein from power — a fact he is careful not to mention in *America at the Crossroads*.

7. Francis Fukuyama, *The End of History and the Last Man* (New York: Simon and Schuster, 1992), 338.

8. Fukuyama himself acknowledged his indebtedness to Marx's vision of historical development and approvingly quotes Ken Jowitt's reflection that "in [September 11's] aftermath, the Bush administration has concluded that Fukuyama's historical timetable is too laissez-faire and not nearly attentive enough to the levers of historical change . . . In this irony of ironies, the Bush administration's identification of regime change as critical to its anti-terrorist policy and integral to its desire for a democratic capitalist world has led to an active 'Leninist' foreign policy in place of Fukuyama's passive 'Marxist' social teleology." Quoted in Fukuyama, *Crossroads*, 55.

9. William F. Buckley, publisher's statement, *National Review*, November 19, 1955.

10. Fukuyama, *Crossroads*, 54.

11. Nietzsche's last man, described in the prologue to *Thus Spoke Zarathustra*, is uncreative and unambitious to the point of seeming sedated. "I tell you," said Zarathustra, "one must still have chaos in oneself to give birth to a dancing star . . . Alas! There comes the time when man will no longer give birth to a star . . . Behold! I show you *the last man*." Friedrich Nietzsche, *Thus Spoke Zarathustra*, trans. Clancy Mar-

tin (New York: Barnes and Noble, 1883, 2005), 13. The apt "couch potato" description of the last man comes from Kathleen Higgins and
Robert Solomon, *What Nietzsche Really Said* (New York: Schocken,
2000), 47.

12. Fukuyama, *Crossroads,* 324.
13. Bell writes, "The central point is that — at first, for the advanced social groups, the intelligentsia and the educated social classes, and later
for the middle class itself — *the legitimations of social behavior passed
from religion to modernist culture.* And with it here was a shift in emphasis from 'character,' which is the unity of moral codes and dis-
ciplined purpose, to an emphasis on 'personality,' which is the en-
hancement of self through the compulsive search for individual
differentiation. In brief, not work but the 'life style' became the source
of satisfaction and criterion for desirable behavior in the society."
Daniel Bell, *The Cultural Contradictions of Capitalism* (New York: Basic Books, 1976), xxiv.
14. Ibid., 13.
15. Ibid., 21.
16. Florida, *Rise,* 207.
17. Fukuyama, *Crossroads,* 326.
18. American conservatives have succeeded, for now, in banning the legal
recognition of same-sex couples in states across the United States. But
they have been so aggressive because they know that time is not on
their side. The percentage of Americans age fifteen and up who agree
with the statement "Society should regard people of the same sex who
live together as being the same as a married couple" increased from 28
percent in 1992 to 45 percent in 2004 — a stunning 17 percent increase
in just fourteen years, according to Environics, Inc., "3SC Values, 1992–
2004."
19. Nietzsche wrote of the possibilities for new ways of living that were
opened up by the decline of traditional religious authority in a famous
passage in *The Gay Science.*

At hearing the news that "the old god is dead," we philosophers and "free spirits" feel illuminated by a new dawn; our heart overflows with gratitude, amaze-
ment, forebodings, expectations — finally, the horizon seems clear again, even
if not bright; finally our ships may set out again, set out to face any danger;
every daring lover of knowledge is allowed again; the sea, *our* sea, lies open
again; maybe there has never been such an "open sea."

Friedrich Nietzsche, *The Gay Science,* trans. Sosefine Nauckhoff (Cambridge, England: Cambridge University Press, 2001), 199.

20. See, for instance, Mark Warren, *Nietzsche and Political Thought* (Cambridge, MA: MIT Press, 1988). For an excellent overview of Nietzsche's multitudes, see especially Alexander Nehamas's landmark *Nietzsche: Life as Literature* (Cambridge, MA: Harvard University Press, 1985), where Nehamas argues that what Nietzsche did was construct a literary figure — "Friedrich Nietzsche!" — through his works, and that if we are to understand Nietzsche's books, we must interpret them as literature. Also see Jacques Derrida, *Spurs: Nietzsche's Styles,* trans. Barbara Harlow (Chicago: University of Chicago Press, 1978); Kathleen Marie Higgins, *Nietzsche's Zarathustra* (Philadelphia: Temple University Press, 1987); and Rüdiger Safranski, *Nietzsche: A Philosophical Biography,* trans. Shelley Frisch (New York: W. W. Norton, 2002).

   And see the first essay, "'Good and Evil,' 'Good and Bad,'" in Friedrich Nietzsche, *The Genealogy of Morals: A Polemic,* trans. Walter Kaufmann and R. J. Hollingdale (New York: Random House, 1969).

21. Christianity (and secular thought) has come a long way since the nineteenth century. Throughout this book, we have described a very different breed of Christian, men like Bono and Rick Warren, who neither valorize meekness nor encourage resentment. They are men who focus the public's attention on *overcoming adversity in this life.* Neither blames AIDS or other kinds of suffering on human sins against God or nature. And neither is particularly meek or humble.

22. Fukuyama, afterword to *End of History.*

23. Peter Schwartz and Doug Randall, "An Abrupt Climate Change Scenario and Its Implications for United States National Security," October 2003, accessible at www.grist.org/pdf/Abrupt+ClimateChange 2003.pdf

24. Paul Berman, *A Tale of Two Utopias: The Political Journey of the Generation of 1968* (New York: W. W. Norton, 1996), 9.

25. Ibid., 294.

26. Economic Policy Institute, "Minimum Wage Facts at a Glance," fact sheet, April 2007.

27. Robert Samuelson, "The Worst of Both Worlds?" *Newsweek,* November 13, 2006, and Samuelson, "The Real Inconvenient Truth," *Newsweek,* July 5, 2006.

28. David Hawkins of the NRDC, telephone interview with M.S., August 12, 2004.

29. See, for instance, Benjamin Friedman, *The Moral Consequences of Economic Growth* (New York: Knopf, 2005), 209–10.

30. Robert Novak, "Bush Withstands G-8 Heat on Kyoto," *Chicago Sun-Times,* July 14, 2005.

31. This point was made by an editorial in *New Scientist,* "Missed Opportunity," July 16, 2005, 3.

32. Geoffrey Lean, "No Thanks, More a Slap in the Face," *Independent,* July 31, 2005.

33. T. R. Reid, *The United States of Europe* (New York: Penguin, 2004), 51–56; Tony Judt, *Postwar* (New York: Penguin, 2005), 153–59; Timothy Garten Ash, *Free World* (New York: Random House, 2004), 30–34.

34. Stewart Brand, quoted in John Tierney, "An Early Environmentalist Embraces 'New Heresies,'" *New York Times,* February 27, 2007.

Bibliography

Ab'Sáber, Aziz. "Problemas da Amazônia brasileira." *Estudos Avançados* 19, no. 53 (2005).

Adams, Michael. *American Backlash*. New York: Penguin, 2005.

Adler, Jonathan H. "Fables of the Cuyahoga: Reconstructing a History of Environmental Protection." *Fordham Environmental Law Journal* 14 (2002).

Amâncio Rodrigues, Osmarino. "O Segundo Assassinato de Chico Mendes." *Cartas da Amazônia*, 1990.

American Environics. "Toward a New Ecological Majority." December 2006. www.americanenvironics.com.

American Lung Association. "Health Disparities in Lung Disease 2006." Fact sheet.

Asper, Alicia. "Latin American Debt Relief: There Is Less Than Meets the Eye." Council on Hemispheric Affairs Analysis, August 2, 2005.

Assis Costa, Francisco de. "Questão agrária e macropolíticas para a Amazônia." *Estudos Avançados* 19, no. 53 (2005).

Atlas, Mark. "Rush to Judgment." *Law and Society Review*, 2001.

Baer, Werner. *The Brazilian Economy*. 4th ed. Westport, CT: Praeger Publishers, 1995.

Bai, Matt. "Who Lost Ohio?" *New York Times Magazine*, November 21, 2004.

Balzhiser, Richard. "The Chinese Energy Outlook." *National Academy of Engineering* 28, no. 2 (Summer 1998).

Banks, James, et al. "Disease and Disadvantage in the United States and in England." *Journal of the American Medical Association*, May 3, 2006.

Becker, Bertha K. "Amazônia: Desenvolvimento, Governabilidade e Soberania, Documento para o IPEA [Instituto de Pesquisa Econômica Aplicada]:

O Estado da Nação." In *Brasil: o estado de uma nação,* edited by Fernando Rezende and Paulo Tafner. São Paulo: Instituto de Pesquisa Econômica Aplicada, 2005.

———. "Geopolítica da Amazônia." *Estudos Avançados* 19, no. 53 (2005).

Bell, Daniel. *The Cultural Contradictions of Capitalism.* New York: Basic Books, 1976.

Bennett, Jane. *The Enchantment of Modern Life.* Princeton, NJ: Princeton University Press, 2001.

Berman, Paul. *A Tale of Two Utopias: The Political Journey of the Generation of 1968.* New York: W. W. Norton, 1996.

Bernard, Alain, Sergey Paltsev, John M. Reilly, Marc Vielle, and Laurent Viguier. "Russia's Role in the Kyoto Protocol." *MIT Joint Program on the Science and Policy of Global Change,* report no. 98 (June 2003).

Berry, Thomas. *The Dream of the Earth.* San Francisco: Sierra Club Books, 1990.

Botkin, Daniel. *Discordant Harmonies.* New York: Oxford University Press, 1990.

Branch, Taylor. *Parting the Waters.* New York: Touchstone, 1998.

Bryant, Mary. "Unequal Justice? Lies, Damn Lies, and Statistics Revisited." *SONREEL News,* September–October 1993, 3–4.

Buckley, William F., Jr. Publisher's statement. *National Review,* November 19, 1955.

Bullard, Robert D., ed. *Confronting Environmental Racism: Voices from the Grassroots.* Boston: South End Press, 1993.

Bullard, Robert D. "Environmental Justice in the Twenty-first Century." In *The Quest for Environmental Justice,* edited by Robert D. Bullard. San Francisco: Sierra Club Books, 2005.

Carson, Rachel. *Silent Spring.* New York: Houghton Mifflin, 1962.

Carvalho, G., A. C. Barros, P. Moutinho, and D. Nepstad. "Sensitive Development Could Protect Amazonia Instead of Destroying It." *Nature* 409, no. 131 (2001).

Centers for Disease Control and Prevention. "Asthma Mortality and Hospitalization Among Children and Young Adults — United States, 1980–1993." *Morbidity and Mortality Weekly Report* 45, no. 17: 350–53.

———. "Cigarette Smoking Among Adults — United States, 2002." *Morbidity and Mortality Weekly Report* 53, no. 20: 427–31.

———. "Percentage of Overweight Among Children and Adolescents, 1999–2002."

Chaloupka, William. "Jagged Terrain: Cronon, Soulé, and the Struggle over

Nature and Deconstruction in Environmental Theory." *Strategies* 13, no. 1 (2000).

Cizik, Richard. "To the Point: Evangelicals and the Environment." *Warren Olney Show.* Public Radio International, February 9, 2005.

Clapp, Phil. "Over Our Dead Bodies: Green Leaders Say Rumors of Environmentalism's Death Are Greatly Exaggerated." Interview by Amanda Griscom Little, January 13, 2005. Accessible at www.grist.org.

Clapp, Richard. "Cancer and the Environment." In *Life Support: The Environment and Human Health,* edited by Michael McCally. Cambridge, MA: MIT Press, 2002.

Cole, Luke, and Sheila Foster. *From the Ground Up.* New York: New York University Press, 2001.

Commoner, Barry. *The Closing Circle.* New York: Knopf, 1971.

Coutinho, Leonardo. "A floresta dá dinheiro." *Veja,* August 22, 2002.

———. "Fiscal, espécie rara." *Veja,* January 28, 2004.

———. "As 7 Pragas da Amazônia." *Veja,* October 12, 2005.

Crichton, Michael. "Environmentalism as Religion." Commonwealth Club, San Francisco, September 15, 2003.

———. *State of Fear.* New York: Harper, 2004.

Critser, Greg. *Fat Land: How Americans Became the Fattest People in the World.* Boston: Houghton Mifflin, 2003.

Davidson, Pamela R. "Risky Business? Relying on Empirical Studies to Assess Environmental Justice." In *Our Backyard: A Quest for Environmental Justice,* edited by Gerald R. Visgilio and Diana M. Whitelaw. Oxford, England: Rowman and Littlefield, 2003.

Derrida, Jacques. *Spurs: Nietzsche's Styles.* Translated by Barbara Harlow. Chicago: University of Chicago Press, 1978.

Diamond, Jared. *Collapse.* New York: Viking, 2005.

Doll, R. "Epidemiological Evidence of the Effects of Behavior and the Environment on the Risk of Human Cancer." *Recent Results in Cancer Research* 154 (1998): 3–21.

Doll, R., and R. Peto. "The Causes of Cancer: Quantitative Estimates of Avoidable Risks of Cancer in the United States Today." *Journal of the National Cancer Institute* 66 (1981): 1191–1308.

Dongfang, Han. "Chinese Labor Struggles." *New Left Review,* July–August 2005.

Dowie, Mark. *Losing Ground.* Cambridge, MA: MIT Press, 1995.

Dubos, René. *So Human an Animal.* New York: Scribner, 1968.

Dupree, Sherry. Interview. *The Tavis Smiley Show.* National Public Radio, August 28, 2003.

Easterbrook, Gregg. *The Progress Paradox.* New York: Random House, 2003.

Eckman, Paul. *Emotions Revealed.* New York: Henry Holt, 2003.

Economic Policy Institute. "Minimum Wage Facts at a Glance." Fact sheet. Accessible at www.epinet.org/content.cfm/issueguides_minwage_min wagefacts.

Economist. "Selling Hot Air." September 9, 2006.

Eliot, T. S. "Quartet No. 1: Burnt Norton." In *Four Quartets.* 1943.

Emerson, Ralph Waldo. "Self-Reliance." In *The Essential Writings of Ralph Waldo Emerson,* edited by Brooks Atkinson. New York: Random House, 2000.

Environics Research Group. "Socio-Cultural Trends: 3SC 1992–2004."

Environmental Protection Agency Office of Transportation and Air Quality. "Light-Duty Automotive Technology and Fuel Economy Trends, 1975 through 2006." Fact sheet. Accessible at www.epa.gov/otaq/cert/mpg/fetrends/420r06011.pdf.

Evans/McDonough Company. Public-opinion survey of 400 likely voters in Pennsylvania. April 22–24, 2003.

Faber, Daniel, and Deborah McCarthy. "Green of Another Color: Building Effective Partnerships Between Foundations and the Environmental Justice Movement." Nonprofit Sector Research Fund, April 10, 2001.

Fearnside, P. M., et al. "A Delicate Balance in Amazonia." *Science,* February 18, 2005.

Feyerabend, Paul. *Conquest of Abundance: A Tale of Abstraction Versus the Richness of Being.* Chicago: University of Chicago Press, 1999.

Florida, Richard. *The Rise of the Creative Class.* New York: Perseus, 2002.

Folha de São Paulo. "Americano Thomoas Lovejoy, pioneiro na pesquisa da Amazônia, diz que transformar árvores em CO_2 afeta soberqania." June 23, 2002.

Foreman, Christopher H., Jr. *The Promise and Peril of Environmental Justice.* Washington, DC: Brookings Institution, 1998.

Frank, Thomas. *What's the Matter with Kansas?: How Conservatives Won the Heart of America.* New York: Henry Holt, 2004.

Friedman, Benjamin. *The Moral Consequences of Economic Growth.* New York: Knopf, 2005.

Friend Harding, Susan. *The Book of Jerry Falwell: Fundamentalist Language and Politics.* Princeton, NJ: Princeton University Press, 2000.

Fukuyama, Francis. *America at the Crossroads: Democracy, Power, and the Neoconservative Legacy.* New Haven, CT: Yale University Press, 2006.

———. *The End of History and the Last Man.* New York: Simon and Schuster, 1992, 2006.

———. "Letter to President Bill Clinton." January 26, 1998.

———. "Letter to President George W. Bush." September 20, 2001.

Garten Ash, Timothy. *Free World.* New York: Random House, 2004.

Gilbert, Daniel. *Stumbling on Happiness.* New York: Knopf, 2006.

Gladwell, Malcolm. "The Cellular Church." *The New Yorker,* September 12, 2005.

Goldberg, Michelle. *Kingdom Coming.* New York: W. W. Norton, 2006.

Gore, Al. *Earth in the Balance.* New York: Houghton Mifflin, 1992.

———. *An Inconvenient Truth.* Emmaus, PA: Rodale, 2006.

Gottlieb, Robert. *Forcing the Spring.* Washington, DC: Island Press, 1993.

Granovetter, Mark. "The Strength of Weak Ties." *American Journal of Sociology,* May 1973, 1360–80.

Gray, John. *Straw Dogs.* London: Granta, 2002.

Hacker, Jacob. *The Great Risk Shift.* New York: Oxford University Press, 2006.

Haidt, Jonathan. *The Happiness Hypothesis.* New York: Basic Books, 2006.

Harvard Center for Cancer Prevention. "Harvard Report on Cancer Prevention." *Cancer Causes and Control* (1996): 7.

Hawken, Paul, Amory Lovins, and L. Hunter Lovins. *Natural Capitalism: Creating the Next Industrial Revolution.* New York: Little, Brown, 1999.

Hays, Samuel P. *A History of Environmental Politics Since 1945.* Pittsburgh: University of Pittsburgh Press, 2000.

Hecht, Susanna, and Alexander Cockburn. *The Fate of the Forest: Developers, Destroyers and Defenders of the Amazon.* New York: Harper Perennial, 1990.

Heidegger, Martin. *Being and Time.* Albany: State University of New York, 1996.

Higgins, Kathleen Marie. *Nietzsche's Zarathustra.* Philadelphia: Temple University Press, 1987.

Higgins, Kathleen, and Robert Solomon. *What Nietzsche Really Said.* New York: Schocken, 2000.

Hill, Julia Butterfly. *The Legacy of Luna.* New York: Harper San Francisco, 2000.

Hoag, Blossom. "The Massachusetts Chapter's Position on the Cape Cod Wind Farm." *Massachusetts Sierran,* Summer 2005.

Human Rights Watch. "China: Harsh Sentences for Labor Activists." May 10, 2003.

IBOPE. Public-opinion survey of 2,000 Brazilians. April 8–13, 2005.

IMDB (Internet Movie Database), imdb.com/title/tt0497116/taglines.

Inglehart, Ronald. *Modernization and Postmodernization.* Princeton, NJ: Princeton University Press, 1997.

Inglehart, Ronald, and Christian Welzel. *Modernization, Cultural Change, and Democracy.* Cambridge, England: Cambridge University Press, 2005.

Instituto Nacional de Pesquisas Espaciais. "Deforestation Estimates for the Brazilian Amazon." Sao José dos Campos, Brazil, 2000.

IPAM. "Avança Brasil, Os Custos Ambientais para Amazônia." 2000.

Jost, John T., et al. "Political Conservatism as Motivated Social Cognition." *Psychological Bulletin* 129, no. 3 (2003): 339–75.

Judt, Tony. *Postwar.* New York: Penguin, 2005.

Jung, C. G. *Mysterium Coniunctionis: An Inquiry into the Separation and Synthesis of Psychic Opposites in Alchemy.* New York: Pantheon, 1963.

Kahn, Jeremy. "Will These People Swing?" *Fortune,* March 22, 2004.

Kaplun, Alex. "EPA Finds Elevated Cancer Rates in Urban Areas." *Greenwire,* February 23, 2006.

Kasser, Jeffrey L. *The Philosophy of Science: Kuhn and the Challenge of History.* Lecture on DVD. Teaching Company, 2006.

Kennedy, Robert F., Jr. "Wind Resistance." *The Connection.* WBUR, October 7, 2002.

Kierkegaard, Søren. *The Journals of Søren Kierkegaard.* Edited by Alexander Dru. Oxford, England: Oxford University Press, 1938.

King, Martin Luther, Jr. "I Have a Dream." Speech at the March on Washington, August 28, 1963.

Kingstone, Steve. "UN Highlights Brazil Gun Crisis." *BBC News,* June 27, 2005.

Knippers Brack, Jan. *United States Penetration of Brazil.* Philadelphia: University of Pennsylvania Press, 1977.

Kramnick, Isaac. Introduction to *Democracy in America,* by Alexis de Tocqueville. New York: Penguin, 2003.

Kuhn, Thomas. *The Structure of Scientific Revolutions.* Chicago: Lawrence Van Gelder, 1962.

Lane, Robert. *The Loss of Happiness in Market Democracies.* New Haven, CT: Yale University Press, 2000.

Latour, Bruno. *The Politics of Nature: How to Bring the Sciences into Democracy.* Translated by Catherine Porter. Cambridge, MA: Harvard University Press, 2004.

Laurance, W. F., A.K.M. Albernaz, P. Fearnside, H. L. Vasconcelos, and L. V. Ferreira. "The Future of the Brazilian Amazon." *Science,* January 19, 2001.

Laurance, W. F., et al. "Letters." *Science,* February 1, 2001.

———. "Letters." *Science,* May 31, 2001.

Lavelle, Marianne, and Marcia Coyle. "Unequal Protection." *National Law Journal,* September 21, 1992, S1–S2.

Levenstein, Harvey. *A Paradox of Plenty: A Social History of Eating in Modern America.* Berkeley: University of California Press, 2003.

Levy, Jonathan, and John D. Spengler. "Estimated Public Health Impacts of Criteria Pollutant Air Emissions from the Salem Harbor and Brayton Point Power Plants." Harvard University School of Public Health, May 2000.

Leydet, François. *The Last Redwoods and the Parkland of Redwood Creek.* San Francisco: Sierra Club Books, 1963.

MacQueen, Kim. "Roads to Ruin." *FSU Research in Review,* Summer 1994.

Mann, Charles C. *1491: New Revelations of the Americas Before Columbus.* New York: Knopf, 2005.

Marcuse, Herbert. *One-Dimensional Man.* New York: Beacon, 1964, 1991.

Maslow, A. H. "A Theory of Human Motivation." *Psychological Review* 50 (1943): 370–96.

———. *The Farther Reaches of Human Nature.* New York: Viking, 1971.

———. *Toward a Psychology of Being.* Foreword by Richard Lowry. New York: John Wiley and Sons, 1999.

McKibben, Bill. "Changing the Climate: Why a New Approach to Global Warming Would Make for a Better Politics — and Planet." *American Prospect,* October 2005, A10.

McPherson, Miller, Lynn Smith-Lovin, and Matthew E. Brashears. "Social Isolation in America: Changes in Core Discussion Networks over Two Decades." *American Sociological Review,* June 2006.

Menand, Louis. *The Metaphysical Club.* New York: Farrar, Straus, and Giroux, 2001.

Mount Sinai School of Medicine. "Mount Sinai Researchers Report Changes in Home Environment Can Reduce Asthma Symptoms in Inner-City Children." Press release summary of *New England Journal of Medicine* article, September 8, 2004.

National Oceanic and Atmospheric Administration. "Buzzards Bay Oil Spill in Massachusetts." Fact sheet, May 2003.

Nehamas, Alexander. *Nietzsche: Life as Literature.* Cambridge, MA: Harvard University Press, 1985.

Nelson, Nancy. "The Majority of Cancers Are Linked to the Environment." *National Cancer Institute Benchmarks* 4, no. 3 (June 17, 2004).

Nepstad, D., O. Almeida, J. Carter, M. C. Vera Diaz, D. McGrath, C. Stickler, P. Pacheco, and D. Kaimowitz. "Globalization of the Amazon Soy and Beef Industries: Opportunities for Conservation." *Conservation Biology* 20, no. 6 (December 2006): 1595–1603.

New Scientist. Editorial: "Missed Opportunity." July 16, 2005.

Nicholas Institute of Environmental Policy Solutions and Peter Hart Research and Public Opinion Strategies. Public-opinion survey of eight hundred voters. August 25–28, 2005.

Nichols, Albert L. "Risk-Based Priorities and Environmental Justice." In *Worst Things First?,* edited by Adam M. Finkel and Dominic Golding. Washington, DC: Resources for the Future, 1995.

Nietzsche, Friedrich. *Ecce Homo.* Translated by Walter Kaufmann. New York: Vintage, 1967.

———. *The Gay Science.* Translated by Josefine Nauckhoff. Cambridge, England: Cambridge University Press, 2001.

———. "'Good and Evil,' 'Good and Bad.'" In *The Genealogy of Morals: A Polemic,* translated by Walter Kaufmann and R. J. Hollingdale. New York: Random House, 1969.

———. *Human, All Too Human.* Lincoln: University of Nebraska Press, 1984.

———. *Thus Spoke Zarathustra.* Translated by Clancy Martin. New York: Barnes and Noble Books, 2005.

Notes from a Personal War. Documentary film on DVD of *City of God,* 2001.

Page, Joseph A. *The Brazilians.* Reading, MA: Perseus, 1995.

Parker, Phyllis. *Brazil and the Quiet Intervention.* Austin: University of Texas Press, 1964, 1979.

Pastors.com. "Rick Warren, Jerry Falwell Meet at 'Purpose-Driven' Common Ground with 13,000." 2003.

People of Color Environmental Leadership Summit. "Principles of Environmental Justice." Adopted October 1991.

Pew Research Center for the People and the Press. "Partisanship Drives Opinion: Little Consensus on Global Warming." Public-opinion survey of 1,501 adults. June 14–19, 2006.

Planck, Max. *Scientific Autobiography and Other Papers.* New York: Philosophical Library, 1949.

Pope, Carl. "There Is Something Different About Global Warming." Response to "The Death of Environmentalism."

Proctor, Jim. "Environment After Nature," 2007. Accessible at www.lclark.edu/~jproctor/pdf/NV2007Draft.pdf.

Putnam, Robert D. *Bowling Alone.* New York: Simon and Schuster, 2000.

Putnam, Robert, and Lewis Feldstein with Don Cohen. *Better Together: Restoring the American Community.* New York: Simon and Schuster, 2003.

Reid, T. R. *The United States of Europe.* New York: Penguin, 2004.

Revkin, Andrew. *The Burning Season.* Washington, DC: Island Press, 1990, 2004.

Rhodes, L., C. M. Bailey, and J. E. Moorman. "Asthma Prevalence and Control Characteristics by Race/Ethnicity — United States, 2002." *Morbidity and Mortality Weekly Report,* February 27, 2004, 145–48.

Rogers, Carl. *On Becoming a Person.* Boston: Houghton Mifflin, 1961, 1995.

Rorty, Richard. *Achieving Our Country.* Cambridge, MA: Harvard University Press, 1998.

———. *Contingency, Irony, and Solidarity.* Cambridge, England: Cambridge University Press, 1989.

Rose, Steven. "What's Not the Matter with the Middle Class?" *American Prospect,* September 4, 2006.

Safranski, Rüdiger. *Nietzsche: A Philosophical Biography.* Translated by Shelley Frisch. New York: W. W. Norton, 2002.

Samuelson, Robert. "The Real Inconvenient Truth." *Newsweek,* July 5, 2006.

———. "The Worst of Both Worlds?" *Newsweek,* November 13, 2006.

Sancton, Thomas A. "What on Earth Are We Doing?" *Time,* January 2, 1989.

Santilli, Márcio, Paulo Moutinho, Stephan Schwartzman, Daniel Nepstad, Lisa Curran, and Carlos Nobre. "Tropical Deforestation and the Kyoto Protocol: An Editorial Essay." *Climatic Change,* August 2005, 267–76.

Schaeffer, R., and R.L.V. Rodrigues. "Underlying Causes of Deforestation." *Science,* February 18, 2005.

Schmink, Marianne, and Charles H. Wood. *Contested Frontiers in Amazonia.* New York: Columbia University Press, 1992.

Schor, Juliet. *The Overspent American.* New York: Basic Books, 1998.

Schwartz, Peter, and Doug Randall. "An Abrupt Climate Change Scenario and Its Implications for United States National Security," October 2003. Accessible at www.grist.org/pdf/AbruptClimateChange2003.pdf.

Schwartzman, Stephen, and Robert Bonnie. "Letters." *Science,* May 31, 2001.

Seligman, Martin. *Authentic Happiness.* New York: Simon and Schuster, 2002.

Serrill, Michael S., and Ian McCluskey. "Unholy Confession." *Time,* May 13, 1996.

Skidmore, Thomas. *Brazil: Five Centuries of Change.* New York: Oxford University Press, 1999.

Sociovision. "The Drivers of Change in a Cross-Cultural World." *GlocalConsult,* 2002.

Southwest Network for Environmental and Economic Justice. Letter to environmental groups. May 20, 1990.

Southwest Organizing Project. Letter to environmental groups. March 15, 1990.

Suskind, Ron. "Without a Doubt." *New York Times Magazine,* October 17, 2004.

Terborgh, John. *Requiem for Nature.* Washington, DC: Island Press, 1999.

Tocqueville, Alexis de. *Democracy in America.* Translated by Gerald Bevan. New York: Penguin, 2003.

Tomasky, Michael. "Party in Search of a Notion." *American Prospect,* May 2006.

Unger, Craig. "American Rapture." *Vanity Fair,* December 2005.

United Church of Christ. Commission on Racial Justice. "Toxic Wastes and Race in the United States: A National Report on the Racial and Socio-Economic Characteristics of Communities with Hazardous Waste Sites." 1987.

United Nations. Climate Change Secretariat. "2006 UNFCCC Greenhouse Gas Data Report Points to Rising Emission Trends." October 30, 2006.

United Nations. Convention on the Prevention and Punishment of the Crime of Genocide. December 9, 1948.

United Nations Framework Convention on Climate Change. "Changes in GHG Emissions from 1990 to 2004 for Annex I Parties."

U.S. Department of Health and Human Services. "Overweight and Obesity Threatens U.S. Health Gains." Fact sheet. December 13, 2001. Accessible at www.surgeongeneral.gov/news/pressreleases/pr_obesity.pdf.

U.S. Energy Information Administration. "China: Environmental Issues." Fact sheet.

———. "Petroleum Products Supplied by Type, 1949–2005." Fact sheet.

U.S. Environmental Protection Agency. "Unfinished Business: A Comparative Assessment of Environmental Problems. Appendix I: Report of the Cancer Risk Work Group." Washington, DC, February 1987.

Vanden, Harry. "Brazil's Landless Hold Their Ground." *NACLA Report on the Americas,* March 1, 2005.

Veja. "A Amazônia será ocupada: entrevisa com David McGrath." November 12, 2003.

Waal, Frans de. "Talking Primates with Dr. Frans de Waal." *92Y Blog,* August 25, 2005. http://blog.92y.org/index.php/item/dr_frans_de_waal/.

Warren, Mark. *Nietzsche and Political Thought.* Cambridge, MA: MIT Press, 1988.

West Harlem Environmental Action, Inc. "Harlem Asthma Study Confirms WE ACT's Claim: If You Live Uptown, Breathe at Your Own Risk." Press release, April 23, 2003.

Wilson, Edward O. *The Future of Life.* New York: Random House, 2002.

Wilson, Timothy. *Strangers to Ourselves.* Cambridge, MA: Harvard University Press, 2002.

World Resources Institute. "Country Profiles."

Younge, Gary. "I Have a Dream." *Guardian,* August 21, 2003.

Index